1993

Conservation Biology
and the Black-Footed Ferret

Conservation Biology and the Black-Footed Ferret

Edited by Ulysses S. Seal

E. TOM THORNE

MICHAEL A. BOGAN

STANLEY H. ANDERSON

Yale University Press
New Haven and London

Published with assistance from the foundation
established in memory of William McKean Brown.

Designed by Jo Aerne and set in Baskerville type
with Futura for display by The Composing Room of
Michigan, Inc.
Printed in the United States of America by
Thomson-Shore, Inc., Dexter, Michigan.

Library of Congress Cataloging-in-Publication Data
Conservation biology and the black-footed ferret /
edited by Ulysses S. Seal . . . [et al.].
    p. cm.
    Based on papers from a meeting.
    Bibliography: p.
    Includes index.
    ISBN 0-300-04123-3 (alk. paper)
    1. Black-footed ferret. 2. Wildlife conservation.
I. Seal, Ulysses S.
QL737.C25C67 1989   88-37644
639.9'7974447—dc19   CIP

The paper in this book meets the guidelines for
permanence and durability of the Committee on
Production Guidelines for Book Longevity of the
Council on Library Resources.

10 9 8 7 6 5 4 3 2 1

Contents

living ferrets are found in locations remote from Meeteetse. Chromosomal analysis will be useful in assessing whether this degree of genetic divergence has occurred.

Reproduction by all of the captive ferrets in order to produce surviving, reproductive offspring is essential if the captive population is to retain the genetic diversity present in the founder stock. The failure to produce offspring at Sybille in the spring of 1986, failure of the South Dakota black-footed ferrets to produce surviving offspring, and the reputed difficulties of breeding wild-caught mustelids placed a high priority on review of current knowledge of the reproductive biology of mustelids and its manipulation.

The reproductive patterns of male and female mustelid species, based upon adequate data from 19 of the 64 species and fragmentary information from 19 others, are diverse, Mead reports. Female patterns include the occurrence of mono- and polyestrous cycles, seasonality, single and multiple litters in a year, and patterns of delayed implantation with embryonic diapause. Black-footed ferrets may be seasonally monoestrous with minimal delay of implantation, and the reproductive lifespan may be no more than 5 to 6 years, thus limiting the opportunities to achieve successful reproduction in a captive colony. Studies of the female reproductive biology of the closely related domestic ferret and mink, described by Murphy, have allowed induction of folliculogenesis and ovulation followed by successful breeding in these species. Such techniques may be necessary for female black-footed ferrets that fail to reproduce naturally.

Synchronization of reproduction to a particular time of year is characteristic of nearly all mustelid species and is easy to rationalize in terms of resource availability for rearing young. Less easy has been the unraveling of the neuroendocrine pathways by which this feat is accomplished. Herbert notes that photoperiod or alterations in day length are the primary proximate cue for timing reproductive events in mustelids. His lucid discussion of the different types of cycle in mustelids, the events in the female cycle subject to control, the role of the pineal and melatonin, and the effects of photoperiod manipulation provide a basis for management of light cycles and timing of reproduction in captive colonies of ferrets. Female ferrets could potentially produce 2 litters per year if increasing or achieving production from founders becomes necessary. Males may be more difficult to manipulate, but collection and cryopreservation of semen might avoid this difficulty.

Because little work had been done on the reproductive biology of male mustelids, such studies were initiated on domestic ferrets as a model for the black-footed ferret. Atherton and colleagues report on abrupt decline in numbers because of disease emphasizes the vulnerability of such populations to random environmental events that can lead to extinction.

An understanding and application of the genetics and demography of small populations are essential to meet the challenge of conserving a species reduced to a small fraction of its original numbers in a single surviving population. The risks of allowing a small remnant population to undergo natural processes in the wild may be evaluated by application of small population biology theory and population viability analysis. Small populations are vulnerable to genetic uncertainty through drift and inbreeding, to demographic uncertainty through fluctuations in age and sex composition and in reproduction, and to environmental uncertainty. These uncertainties can be evaluated in simulation models presented by Brussard and Gilpin. Given the proven vulnerability of the black-footed ferret to one environmental uncertainty—canine distemper—they suggest a metapopulation structure for a recovery program sustained by active management through recolonization of extinct populations and occasional genetic exchange. They introduce the important concept of effective population size and suggest a total target population of 2,000 or more animals for a recovery program based upon wild populations.

Simulations by Lacy and Clark, based on several scenarios of the genetic and demographic history of the Meeteetse population, suggest that it may have lost 60% of the heterozygosity present in the population in 1930. They also suggest the rapid expansion of a captive population to minimize further genetic loss and show the genetic value of adding additional animals from the wild. Simulations of the probability of extinction of small black-footed ferret populations by Harris, Clark, and Shaffer, based upon estimates of demographic and environmental fluctuations or stochasticity, indicate that populations of more than 100 ferrets are needed to reduce the extinction probability below 5% per 100 years. The authors suggest the need for numerous independent populations in the wild to protect against the vagaries of an uncertain environment.

The first concern of the captive population is demographic. It is essential that breeding be established and that the population expand. As this occurs, Ballou observes, it will then be possible to attend to any effects of inbreeding and to be more selective in matings. Identifying individuals and maintaining pedigree records will be critical. The possibility of outbreeding depression or reduction in fitness from breeding distantly related individuals will need to be considered only if additional vival plans for captive populations of endangered species in zoos has identified common biological themes of management, small population genetics, demography, population viability analysis, small population biology, reproductive biology, and taxonomic status that apply to crisis management for individual species survival whether in captivity or in the wild.

The near extinction of the black-footed ferret, like that of the red wolf (*Canis rufus gregoryi*), represents the loss of a species that was once widely distributed on the North American mainland. This scenario is in sharp contrast to the loss of island-endemic species, which represent about 75% of the well-documented mammalian extinctions of the past 400 years, as Flesness observes. The black-footed ferret and red wolf provide highly visible evidence of an increasing rate of species extinction largely as the result of human activities. The black-footed ferret may have been particularly vulnerable because of its fatal lack of resistance to canine distemper and its dependence on a single prey—the prairie dog, which is systematically poisoned as a matter of public and private land-management policy. Prairie dog towns were once distributed over 40 million hectares; it now appears difficult to allocate even 3,000- to 15,000-hectare patches of habitat to support ferrets.

The fossil record, analyzed by Anderson, indicates that black-footed ferrets may have made their appearance in North America about 100,000 years ago. The black-footed ferret is closely related to *Mustela eversmanni*, the steppe polecat which utilizes multiple prey species; they have been suggested to be conspecific. The attribution of species status by Anderson, from morphological comparisons, has been augmented by the molecular studies of O'Brien and colleagues which indicate a genetic distance from *M. eversmanni* appropriate for species distinction in carnivores and mustelids. The black-footed ferrets, however, are depauperate in electrophoretically measured polymorphisms and genetic heterozygosity, perhaps reflecting the results of genetic drift over generations of isolation at Meeteetse—a founder bottleneck. There have been no subspecies of *M. nigripes* named, and Anderson has found no evidence of geographic variation in the samples she has studied—an important point if additional live animals are found in other localities. Such animals could be an important addition to the species gene pool.

The black-footed ferret colony near Meeteetse was small, isolated, and probably oscillated in numbers. The estimated numbers of breeding adults ranged from 15 to 35; the highest total population estimate was about 130 in the fall of 1984; and the colony may have been genetically isolated for 50 years, or 20 to 30 ferret generations. The recent

ULYSSES S. SEAL

Introduction

In 1986, the 18 known living wild-born black-footed ferrets were brought into captivity as the result of analyses indicating that the risk of extinction of the single wild population was greater than the risks of capture and establishment of a reproducing colony in captivity. This decision was not made lightly since captive black-footed ferrets have not previously produced surviving offspring.

Capturing all of the remaining ferrets in order to save the species represents a departure from traditional conservation approaches that focus on protection and enhancement of wild habitat. This judgment is in conflict with traditions in wildlife biology, wildlife management, and ecology that regard wild animals in captivity as subject to rapid domestication, degeneration, and loss of those traits and capabilities necessary for survival in the wild. Desperate circumstances and solid scientific rationale from the emerging discipline of conservation biology led to this decision, against substantial resistance, twice in 1 year in North America—for the California condor and for the black-footed ferret. A policy supporting the use of captive breeding for preservation of endangered species, formulated by the Captive Breeding Specialist Group from the concepts of small population biology, has recently been adopted by the International Union for Conservation of Nature and Natural Resources.

Rescue of a species from extinction, when its numbers are few, is a deceptively difficult goal. How difficult can be seen from the array of disciplines represented by the authors of these chapters. Yet the coverage of problems is incomplete, with little consideration of land-use philosophies, control programs, field studies, epidemiology, nutrition, toxicology, behavior, or release biology. All will be a necessary part of a release program of captive-bred ferrets. The first step, however, is establishment of a captive population—secure genetically and demographically—which retains the maximum amount of genetic diversity present in the available founder stock. The process of developing sur-

ULYSSES S. SEAL, E. TOM THORNE,

MICHAEL A. BOGAN, AND STANLEY H. ANDERSON

Preface

There has been a rapid growth of research and theory in small population biology, single species ecology, reproductive technology, molecular systematics, and captive-breeding biology that might be applied to the program for conservation and recovery of the black-footed ferret since the 1985 workshop on this species. The meeting from which this book emanated was organized to bring together active workers in these disciplines to provide a state-of-the-art survey with emphasis on direct application to the recovery program for the black-footed ferret. Many of the speakers had recently been recruited to do specialized study on the black-footed ferret or closely related species. The results of the meeting have helped guide development of many aspects of the recovery program.

The presentations and discussions clarified certain controversial issues in conservation and wildlife biology, including factors influencing the viability of small wild and captive populations, minimum viable population sizes in wild and captive populations, and the consequences of small founder numbers for recovery of the species. These papers were useful in the decision-making stage of the recovery program and will assist in the return of the species to the wild—the goal of a recovery program.

We are grateful to the many people who helped make the meeting possible and successful, including members of the Wyoming Game and Fish Department, the University of Wyoming, the U.S. Fish and Wildlife Service, the Captive Breeding Specialist Group, and, most important, the speakers and their colleagues.

the collection and cryopreservation of semen from domestic ferrets, noting that at least 1 female has been successfully inseminated with frozen and thawed semen. The details of their work make clear that the potential application of reproductive technology to preservation and population expansion of endangered species will require substantial time and money. Despite many years and millions of dollars spent in developing these technologies in humans, cattle, and a few other domestic species, transferring the techniques to new species has proved difficult. The same is true for the embryo technology described by Wildt and Goodrowe in their review of developments with carnivores and analysis of the potential of this technology for ferrets. Detailed knowledge of the female reproductive cycle and its manipulation is necessary, but they suggest that successful application of these techniques to ferrets is within our grasp if funds and animals are made available. In addition to achievement of reproduction in reluctant females and males and to expansion of living populations from a limited number of founders, there is the potential for long-term cryo-storage of embryos to preserve more of the genetic diversity of species whose wild habitat is currently limited.

Reduction in numbers, vulnerability to unexpected events, and fragmentation of habitat have made intensive management of small wild populations of many vertebrate species necessary for their survival as other than curios. Distinctions between wild and captive management are rapidly blurring, to the dismay of all. It will be necessary, for some species in North America, to manage many separate, small populations both in the wild and in captivity because sufficiently large habitat patches cannot be secured to allow management by benign neglect. It is likely that the black-footed ferret is one such species.

Foose reports that the theory and practice of small population biology is being developed by zoos to manage small dispersed groups of animals; these institutions are actively collaborating to retain genetic diversity and to protect from demographic catastrophe. Species Survival Plans are being constructed to meet this need and are resulting in a level of sophistication in zoo population management rarely applied to wild populations of endangered species. Bridges between the communities of wildlife biologists, ecologists, and zoo biologists are needed. Joint planning and management teams devoted to the conservation of individual species in the wild—using captive populations as an essential component—have recently been formed for several species, including the black-footed ferret and red wolf.

Recovery in the wild is the goal of planning for listed species by the Fish and Wildlife Service under the Endangered Species Act and other legislation. The concept of recovery is not clearly defined in biological terms in the act but includes the removal of threats that contribute to the decline of the species. It also involves, according to Cole, restoration of population numbers in the wild "to a point where it is a viable self-sustaining component of its ecosystem and no longer in need of protection." He notes that more than 400 species in North America have been listed, that 1,000 candidate species warrant listing, and that the rate of listing is about 50 per year. Clearly the task has exceeded the allocated resources. The black-footed ferret, however, has high priority for existing resources.

The 1986 surveys of the black-footed ferrets near Meeteetse, reviewed by Belitsky, indicated 14 animals remaining, 4 of which were brought into the captive colony, leaving 10 known animals as possible additional candidates for capture. A captive colony of black-footed ferrets was established in existing facilities at Sybille; and a new facility has been built at the site described by Thorne, who carries the responsibility for the captive program and for veterinary management of the species in the wild and in captivity. During the 1986 mating season, efforts were made to breed the 2 male and 4 female black-footed ferrets captured in October and November of 1985. Although these attempts were unsuccessful, important aspects of their behavior and management during this time were studied and described by DonCarlos, Miller, and Thorne.

Given the small number of founders and their probable relatedness, captive breeding must be done systematically with careful attention to pedigrees in order to retain the genetic variation present in the captive population and to avoid unnecessary inbreeding. There is the further need to establish population goals for the captive population to provide a resource for the release program that will not endanger the genetic and demographic integrity of the species. Ballou and Oakleaf, authors of a population biology management plan, observe that the effective number of founders is no more than 9 or 10 considering the sex ratio and relationships of all of the known wild-caught animals. Retention of 80 or 90% of the genetic diversity in this small founder population will require rapid expansion to between 500 and 2,000 animals, an effort that implies building the population to a size at which heterozygosity may be more rapidly gained by mutation than lost by genetic drift. These estimates place a minimum size upon the total population needed in the wild to achieve recovery goals.

A purpose of this book was to provide a state-of-the-art assessment and presentation of the science that might be applied to the problem of black-footed ferret survival and recovery. This information can provide decision makers a rational basis in biological science for the choices to be made and risks to be taken to save the black-footed ferret as a species. It was intended to reduce the uncertainties about small population biology, reproductive biology, and captive management of mustelids as applied to black-footed ferrets and to allow other areas of conflict—philosophies, traditions, and politics—to be resolved on their own merits. However, there remain biological uncertainties with their associated risks that must be weighed in this complicated decision-making process. Formal techniques now available to assist this process have only recently been applied to problems of conservation biology. An advantage of formal analysis is that it forces explicit formulation and weighting of parameters being used in the decision-making process. The analysis by Maguire of the capture and release decisions to be made for black-footed ferrets provides a robust general conclusion that capture of all remaining ferrets was probably an optimal strategy. The decision was ultimately made to capture these animals and to incorporate them into the captive-breeding program.

PART ONE

Context and Systematics

1

NATHAN R. FLESNESS

Mammalian Extinction Rates:
Background to the Black-Footed Ferret Drama

Population estimates indicate that the black-footed ferret (*Mustela nigripes*) is having a very close brush with extinction. Current estimates suggest that these animals are closer to extinction than any other mammal species in the world.

Their fate should be seen in a broader context. Species have always been going extinct, but the current rate is far above normal. Human activities have greatly speeded up extinctions. We are significantly reducing the number of life-forms sharing the planet with us.

There are about 4,000 existing mammal species on earth (4,008 in Corbet and Hill 1980; 4,170 in Honacki et al. 1982—both list some extinct species). We can imagine this number as an approximate equilibrium maintained by a very rough balance between extinction and speciation. With such an assumption it is possible to calculate a normal mammalian species extinction rate given one more measurement: average species longevity.

Available estimates for species longevity values come from paleontologists and have recently been discussed by Schopf (1982), Stanley (1985), and Gingerich (1985). Schopf argues that such values have often been overestimated and suggests a typical value of 200,000 years. At a 4,000 mammalian species equilibrium, this would mean an average of one extinction every 50 years, or 0.02 per year.

Stanley (1985) and Gingerich (1985) support a more traditional, longer species duration. Their values range from 1 to several million years; two million will serve as a rough average. For the same equilibrium, this would indicate one mammal species extinction every 500 years, or 0.002 per year.

In either case, maintaining the assumed equilibrium requires a speciation rate that roughly matches the extinction rate. Some researchers (Soule and Wilcox 1980) have expressed concern that speciation rates

3

will slow down for some of the same reasons that extinction rates are presently rising.

Current and Historical Extinction Rates

Written records that bear on mammal extinctions go back to records from the European voyages of discovery and date to about 1600. Retrospective compilation of these records of sightings—then records of failure to sight—permits assembly of an approximate historical extinction record (Allen 1942; Goodwin and Goodwin 1973). Caution is needed because records were probably poorer in the earlier years. In fact, until 1800 most educated Westerners did not think extinction of species was even possible (Rudwick 1972; Grayson 1984).

The worldwide historical record of mammalian species extinctions (Fig. 1.1) lists 58 species lost in the past 400 years. This is an overall rate of 0.15 mammalian species extinctions per year, or 1 about every 7 years. This is 7 to 70 times faster than indicated in fossil records.

During this century, 23 mammalian species extinctions have been recorded. This is 0.27 per year, or 1 every 4 years. These current losses are 13 to 135 times higher than the normal fossil record rate.

The rate of loss of mammal species appears to be accelerating, although with the limited historical data the high correlation of mammal extinctions per century ($r = .85$) is not quite significant ($p = 0.08$). Even without acceleration, if the rate of loss for this century is maintained, each human generation of threescore and ten years will witness the end of 19 mammal species.

Forty-three of the 58 recorded mammalian species extinctions (or 75%) are of island forms. Tabulating subspecies and species extinctions, Diamond (1984) found that most disappearing taxa are subspecies of mainland forms. These two results suggest that we are exterminating island species while eroding mainland species piece by piece. Island populations of most vertebrates are small and have evolved in a world of limited diversity. Encounters with greater diversity (man, his commensals, his pests, and diseases carried by any of these) have often resulted in extinction. Like the black-footed ferret, the Tasmanian thylacine may also have been affected by a distemper virus (Day 1981).

Six of the 58 mammal species lost in recorded history are from the Australian mainland, where combined pressures of introductions, range modification, and hunting probably affected survival. The remaining 9 mammal extinctions were recorded on the larger continents and in the oceans and are thought to have occurred primarily because of

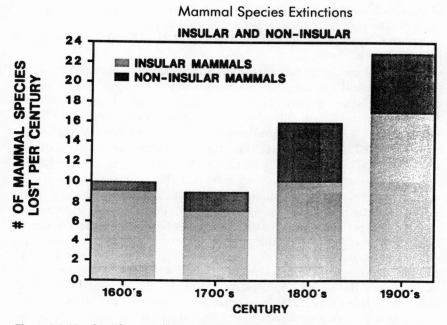

Figure 1.1. Number of mammalian species known to have been lost since 1600 by century. Insular and mainland species are tabulated separately.

hunting or trapping (Day 1981). The potential role of introduced disease is hard to ascertain, however. Black-footed ferrets fit this pattern in the sense that their apparently near-exclusive prey has been spectacularly reduced by intentional human activity—an activity as destructive as hunting the ferret directly.

North American Extinctions

Records show that since 1600 the North American continental region has lost 35 mammal species. By the 1890s, 25 of those species had already been lost—all insular endemics, primarily rodents and bats from Caribbean islands. In the past 90 years, 10 more have been lost, but for the first time extinctions in North America include 2 mammal species that were not restricted to 1 or a few islands.

These 2 species were the sea mink, *Mustela macrodon*, of the New England coast (which vanished about 1894) and the Caribbean Monk seal, *Monachus tropicalis*, originally widespread throughout the Carib-

bean (which was last seen about 1952). Excessive fur trapping and hunting were the probable cause of both extinctions. (Some taxonomic treatments identify the sea mink as a subspecies [Honacki et al. 1982; see also
Corbet and Hill 1980].)

Ignoring the insular endemics, 2 such extinctions in 90 years amounts
to 0.02 mainland species per year. Such a number appears small but is
deceptive. The highest mammalian extinction rate in the past 10 million
years was for the Rancholabrean, when most of North America's mega-
fauna (about 72 land mammal species) were wiped out (Webb 1984;
Graham and Lundelius 1984). The Rancholabrean lasted a few thousand years; using three for "few," this is a loss of 0.024 species per year.
Thus, extinction rates like those achieved this century in mainland
North America can have dramatic effects over the long term.

True insular populations are more likely to go extinct for 2 different
sets of reasons (Temple 1986). One set arises from their inherent small
size and range. Natural or man-made disasters, smaller perturbations
seen as environmental and demographic stochasticity, or occasionally
even inbreeding effects can threaten them. The second set, based on
evolutionary development, consists of insular adaptations that later
make populations vulnerable to introduced species and disturbance.

Humans progressively reduce habitats available to many species of
wildlife—by amounts that in "developed" countries may often reach 95
to 99%: 99% of U.S. prairie is gone (Myers 1979); 90 to 97% of U.S.
prairie dog towns are gone (Forrest et al. 1985; Gates 1973; Thornback
and Jenkins 1982); on a broader scale about 97% of the land area of the
earth is developed or is implicitly planned to be—since only about 3% is
scheduled for protection.

Remaining populations are not just smaller, they are typically quite
discontinuous. Habitat islands are formed, between which migration
may be effectively impossible. Within the islands, population sizes may
be too low for persistence of many forms, especially large animals or
carnivores. Insular demographic risks, mentioned above, operate on
these newly formed islands of habitat.

The population on each small, isolated habitat island may flicker and
die out, while recolonization from other areas may be too improbable to
be of practical importance. There are many small, protected prairie
habitats in the contiguous 48 states, but almost none of them retain
buffalo, and not one retains the original plains grizzly and wolf.

Some rough approximations illustrate the importance of this simul-
taneous reduction and fragmentation of the habitat for black-footed
ferrets. Sophisticated demographic stochastic modeling has been applied to black-footed ferret populations (Harris et al., chap. 6, and

Brussard and Gilpin, chap. 4, this volume), but even informal estimates give perspective.

Suppose that the time span of concern is several decades—roughly the time since the massive elimination of the prairie dog occurred. Given the high mortality rates reported, it is likely that individual black-footed ferret populations must number 50 or so to have a substantial chance of surviving 20 to 30 years.

Ferrets are thought to require about 50 hectares of prairie-dog-town habitat per animal (Forrest et al. 1985), so for the given time span, the estimated minimum of prairie-dog-town habitat must be 2,500 hectares or more to have a reasonable chance of sustaining a ferret population. Meeteetse is 2,886.5 hectares, and the now-vanished population of Mellette County, South Dakota, had about 650 hectares of prairie dog towns (Linder and Hillman 1973, 21).

Such arguments suggest that black-footed ferrets are already extinct in most (smaller) prairie-dog-town complexes, and that additional wild populations persist only in existing larger complexes. An inventory of extant larger prairie dog complexes, and any recoverable history of their poisoning, is needed. The same arguments suggest that if managers must work with smaller prairie dog complexes to sustain possible future ferret populations (Forrest et al. 1985), management will need to be relatively intensive.

On the other hand, the canine distemper epizootics that have caused high mortality in the Patuxent and Sybille captive populations, and in at least the Meeteetse wild population, suggest that isolated populations of black-footed ferrets may have the best chance of surviving. Therefore, any wild populations beyond Meeteetse will do best in the larger but most isolated prairie dog complexes.

Extinction can be a natural event. Yet almost all mammalian extinctions recorded historically have been caused by human activities, not natural events. Available evidence indicates that mammalian extinctions are becoming more and more frequent and currently occur at a rate that matches or exceeds that of the biggest mammalian extinction catastrophe of the past 10 million years.

The black-footed ferret may be added to this sorry list on our watch. If so, it will be because of our intentional reduction of its obligate prey species, and possibly because of our inadvertent introduction of an exotic virus.

It will also be because of a very real lack of perspective. There are 104 million hectares of publicly owned rangelands in the United States (Wolf 1986). Ferret survival requires that tens of thousands of them, containing larger prairie dog complexes (perhaps newly established), be protected for prairie dogs and ferrets. This would require a commitment of perhaps 0.03%. The question is, will a rich country like the United States be willing to dedicate 1/3,000th of existing public rangelands to save a mammal species critically endangered by our range management techniques?

References

Allen, G. M. [1942] 1972. *Extinct and vanishing mammals of the Western hemisphere.* Reprint. New York: Cooper Square Publ.

Corbet, G. B., and J. E. Hill. 1980. *A world list of mammalian species.* Ithaca, New York: Cornell Univ. Press.

Day, D. 1981. *The doomsday book of animals.* New York: Viking Press.

Diamond, J. 1984. Historic extinction: A rosetta stone for understanding prehistoric extinctions. In *Quaternary extinctions,* ed. P. S. Martin and R. G. Klein, 824–62. Tucson: Univ. of Arizona Press.

Forrest, S. C. et al. 1985. Black-footed ferret habitat: Some management and reintroduction considerations. *Wyoming BLM Wildlife Technical Bulletin* 2.

Gates, J. M. 1973. Introduction. In *Black-footed ferret and prairie dog workshop,* ed. R. Linder and C. Hillman. Brookings, S. Dak.: South Dakota State Univ.

Gingerich, P. 1985. Species in the fossil record: concepts, trends, and transitions. *Paleobiology* 11:27–41.

Goodwin, H. A., and J. M. Goodwin. 1973. List of mammals which have become extinct or are possibly extinct since 1600. IUCN Occ. Paper No. 8. Gland, Switzerland: International Union for Conservation of Nature and Natural Resources.

Graham, R. W., and E. L. Lundelius, Jr. 1984. Coevolutionary disequilibrium and Pleistocene extinctions. In *Quaternary extinctions,* ed. P. S. Martin and R. G. Klein, 5–39. Tucson: Univ. of Arizona Press.

Grayson, D. K., 1984. Nineteenth-century explanations of Pleistocene extinctions: A review and analysis. In *Quaternary extinctions,* ed. P. S. Martin and R. G. Klein, 223–49. Tucson: Univ. of Arizona Press.

Honacki, J. H., K. E. Kinman, and J. W. Koeppl, eds. 1982. *Mammal species of the world,* Association of Systematics Collections. Lawrence, Kans.: Allen Press.

Linder, R., and C. Hillman, eds. 1973. *Black-footed ferret and prairie dog workshop.* Brookings, S. Dak.: South Dakota State Univ.

Myers, N. 1979. *The sinking ark.* New York: Pergamon Press.

M. J. S. Rudwick. 1972. *The meaning of fossils.* London: McDonald Press.

Schopf, T. J. M. 1982. A critical assessment of punctuated equilibria: I. Duration of taxa. *Evolution* 36:1144–57.

Soule, M., and B. Wilcox. 1980. Conservation biology: Its scope and its challenge. In *Conservation biology, an evolutionary-ecological perspective,* ed. M. Soule and B. Wilcox, 8. Sunderland, Mass: Sinauer Assoc.

Stanley, S. M. 1985. Rates of evolution. *Paleobiology* 11:13–26.

Temple, S. 1986. Why endemic island birds are so vulnerable to extinction. In *Bird conservation 2,* ed. S. Temple, 3–6. International Council Bird Preservation, U.S. Section. Madison: Univ. of Wisconsin Press.

Thornback, J., and M. Jenkins. 1982. *IUCN mammal red data book,* Part 1. Gland, Switzerland: International Union for the Conservation of Nature and Natural Resources.

Webb, S. D. 1984. Ten million years of mammal extinctions in North America, In *Quaternary extinctions,* ed. P. S. Martin and R. G. Klein, 189–210. Tucson: Univ. of Arizona Press.

Wolf, E. C. 1986. Managing rangelands. In *State of the world 1986,* ed. L. Brown, 62–77. New York: Norton and Co.

2

ELAINE ANDERSON

The Phylogeny of Mustelids and
the Systematics of Ferrets

As a Quaternary mammalogist, I am especially interested in the fossil history of the small carnivores and, in particular, the mustelids. In this chapter I will briefly trace the fossil history of the Mustelidae and discuss the subfamilies of the Mustelidae, the genus *Mustela,* the subgenera of *Mustela,* and especially the subgenus *Putorius,* the ferrets.

Fossil mustelids are a difficult group to work on because their generally small size and fragile bones reduce the chance of fossilization. The incomplete fossil record for most lineages has led to confusion about their true relationships. To compound the problem, many workers have described a specimen as a new genus or species without doing adequate comparative studies. This has sometimes resulted in a genus being misplaced in a subfamily or even family. Recently Schmidt-Kittler (1981) in his study of the middle-ear region of mustelids and procyonids discovered that many of the Oligocene and early Miocene taxa that had been placed in the Mustelidae have a procyonid type of ear structure and actually belong to the musteloid stem group from which procyonids were derived. Such problems reflect the importance of clearly describing early carnivores and mustelids.

The order Carnivora is derived from an ancestral insectivore stock represented by the family Miacidae in the middle Paleocene about 60 million years ago (mya). Miacids are characterized by small size, a long body and tail, and short legs. The P^4 and M_1 were developed as shearing carnassial teeth. Miacids were forest dwellers and many were arboreal (Romer 1966).

In the Eocene and early Oligocene (between 57 and 34 mya) diversification among the miacids led to the modern carnivore families. They are divided into two main groups: the arctoid carnivores (mustelids, procyonids, canids, and ursids) and the aeluroid carnivores (viverrids, hyenids, and felids).

The basal group of the arctoid carnivores is the Mustelidae. The early mustelids are rare and poorly known. There were a number of diverse forms, most of which became extinct. But by the middle Miocene (about 12 mya), there were recognizable martens, weasels, polecats, honey badgers, badgers, skunks, and otters.

Numerous genera and species of Tertiary mustelids have been described, often from fragmentary material—some extinct taxa are known only from a single tooth. As Harrison (1981, 10) noted, "Too often individual variation is interpreted as variation at the specific level." A revision of the Tertiary mustelids is badly needed.

There are a few mustelids that have a good fossil record dating back to the late Miocene. These include *Mellivora*, the honey badger whose fossil history in Africa can be traced from the late Miocene to the Recent without interruption (Hendey 1978). *Plesiogulo*, the probable ancestral wolverine, arose in Asia in the Miocene and spread to Europe and South Africa. By the late Miocene, it had reached western North America, where it has been found at 14 localities and is regarded as an indicator species for the Hemphillian land mammal age (Harrison 1981). *Enhydriodon*, the crab-eating otter, has been found in late Miocene deposits in Spain, Italy, and western North America. A trend towards an increase in the width of the upper carnassial (P^4) and inflation of the tooth cusps to a bulbous condition culminated in *Enhydra*, the extant sea otter that made its first appearance in California about 2 mya (Repenning 1976).

Mustelids have retained many primitive characteristics, including relatively small size, short, stocky limbs, 5 toes on each foot, a long braincase, and a short rostrum. They typically show reduction and sometimes loss of the anterior premolar, the loss of the second and third upper molars, reduction of the second lower molar, and loss of the third, so that mustelids have 1 upper and 2 lower molars. This has been accompanied by a shortening of the facial region including the tooth row; thus, there is rarely a diastema between the canine and first premolar or between the premolars. The carnassial teeth are typically sectorial, but in some groups (badgers, skunks, and sea otters) they have been secondarily modified for crushing. The postglenoid process of the skull partially encloses the glenoid fossa so that little lateral and no rotary jaw movement is possible. Fine fur is characteristic of the family. Mustelids are distributed globally, with the exceptions of Australia, oceanic islands, and Madagascar. They are found in a wide variety of habitats ranging from the arctic tundra to the tropical rain forest. No other carnivore family shows such diversity of adaptive types.

Six subfamilies of the Mustelidae are recognized: the Mustelinae (martens, weasels, mink, and ferrets), the Galictinae (tayras and grisons), the Mellivorinae (honey badgers and wolverine), the Melinae (badgers), the Mephitinae (skunks and stink badger), and the Lutrinae (otters and sea otter). Schmidt-Kittler (1981) points out that all the subfamilies are in need of revision. Only the Mephitinae with the addition of *Mydaus,* the stink badger, form a natural group. The other subfamilies contain taxa that show parallel adaptations to a similar habitat and may not reflect true relationships. There are about 23 extant genera and about 62 extant species in the family (Table 2.1). Many of them are poorly known and are represented in collections by only a few specimens. In addition, nothing is known of their habits or reproduction in the wild. Simpson (1945) listed 45 extinct genera of mustelids. Some of these, however, have been transferred to other families or synonymized with other genera; a number of new genera have been described since his findings.

The Mustelinae is the central subfamily of the Mustelidae, but unfortunately it has been used as a catchall for many of the early, poorly differentiated taxa as well as divergent genera of doubtful affinity, so that determining the earliest true members of the subfamily has been nearly impossible. The Mustelinae probably originated in Eurasia. They do not show the specializations for different modes of life that are characteristic of some of the other subfamilies, and they have retained some early mustelid characteristics (small size, sectorial dentition). They are more strictly carnivorous than the other groups and are efficient predators on small animals. The Mustelinae are found mainly in northern temperate regions but also range southward into South America and Africa. Some workers believe that the South American weasels (subgenus *Grammogale*) are distinct enough to be placed in their own genus and subfamily. The African members, *Ictonyx, Poecilictis,* and *Poecilogale,* are poorly known, and their taxonomic relationships are uncertain. The best-known members of the Mustelinae are the martens (genus *Martes*) and the weasels, mink, and ferrets (genus *Mustela*). *Martes* is distinguished from *Mustela* by larger size and by having 4 premolars in both jaws (*Mustela* has 3).

The genus *Mustela* is divided into 5 subgenera: *Mustela* (weasels), *Lutreola* (European mink), *Vison* (American mink), *Putorius* (ferrets), and *Grammogale* (South American weasels) (Table 2.2).

Two distinct lineages of weasels arose in central Europe in the Pliocene (about 4 mya). The first, the ermine lineage, started with *M. plioerminea;* it was ancestral to *M. palerminea* (early middle Pleistocene) and gave rise to *M. erminea,* the extant ermine or stoat. By the late middle Pleistocene (about 1.2 mya), it had spread across Eurasia to North Amer-

Table 2.1. Subfamilies of the Mustelidae

Comparison	Subfamily					
	Mustelinae	Galictinae	Mellivorinae	Melinae	Mephitinae	Lutrinae
Representative genera	*Martes* *Mustela* *Ictonyx* *Vormela* *Poecilictis* *Poecilogale*	*Eira* *Galictis* *Lyncodon* **Trigonictis* **Enhydrictis* **Pannomictis*	*Gulo* *Mellivora* **Plesiogulo* **Promellivora*	*Taxidea* *Meles* *Arctonyx* *Melogale* **Pliotaxidea*	*Mephitis* *Spilogale* *Conepatus* *Mydaus* **Brachyprotoma* *Buisnictis*	*Lutra* *Aonyx* *Enhydra* *Pteroneura* **Enhydriodon* **Satherium*
Geologic range	? Oligocene–Recent	Pliocene–Recent	Late Miocene–Recent	Late Miocene–Recent	Late Miocene–Recent	Mid-Miocene–Recent
Geographic range	North America, Central America, northern South America, Eurasia, Southeast Asia, Africa	Central and South America; extinct forms in Eurasia, North America	Holarctic, Africa, Asia	North America, Mexico, Europe, Asia, Malay Peninsula, Philippines	North America, Central America, South America	Eurasia, Africa, North America, Central America, South America
Habitat	Nearly all terrestrial habitats	Forests, open area	Taiga, tundra, tropical forests	Plains, wooded regions, tropical forests	Woods, plains, deserts	Inland waterways, sluggish streams, N. Pacific coast
Remarks	Least specialized, highly carnivorous	Poorly known	Exceptionally strong for their size	Fossorial, omnivorous	Omnivorous, well-developed anal glands	Aquatic, teeth specialized for crushing

Note: * = extinct genus.

Table 2.2. Comparisons between the subgenera of *Mustela*

Comparison	Subgenus				
	Mustela	*Lutreola*	*Vison*	*Putorius*	*Grammogale*
Geologic range	Pliocene–Recent	Recent	Mid-Pleistocene–Recent	Late Pliocene–Recent	Recent
Geographic range	North America to northern South America, Eurasia	Europe, Siberia, Southeast Asia	North America except arid areas, Introd. into Europe & Siberia	Eurasia, Great Plains of North America	Amazon basin in Ecuador, Columbia, Brazil, Peru
Habitat	Nearly all terrestrial habitats	Small woodland streams, marshes	Along inland streams, marshes	Steppes, prairies, meadows, open forest	Humid riparian forests
External characteristics	Brown above, whitish below; northern species turn white in winter	Brown with white circumoral band; webbed feet	Rich dark brown; white chin patch, webbed feet	Yellowish buff to dark brown; some have dark facial markings and feet	Blackish brown, ventral stripe, webbed feet
Size[a]	M: TL 180–390 T 30–152 Wt. 39–340 g F: TL 165–355 T 25–127 Wt. 38–198 g	M: TL 360–420 T 150–180 Wt. 800–1,005 g F: TL 310–380 T 120–145 Wt. 505–700 g	M: TL 510–570 T 180–230 Wt. 680–1,360 g F: TL 430–560 T 130–200 Wt. 565–1,089 g	M: TL 450–740 T 80–190 Wt. 500–2,050 g F: TL 360–700 T 70–180 Wt. 645–1,360 g	M: TL 324–548 T 210–234

14

Food	Small mammals, birds	Insects, crayfish, mollusks, fish, frogs, birds, small mammals	Aquatic mammals, birds, frogs, fish, crayfish	Rodents, rabbits, birds, frogs	Small animals, aquatic life
Habits	Mostly nocturnal	Semiaquatic	Nocturnal, solitary dens along streams	Mostly nocturnal, utilize rodent burrows	Semiaquatic, good swimmers and climbers
Reproduction	Some species have long delayed implantation; 4–8 young born April–May	No delayed implantation. Gestation period 35–43 days; 2–7 young born April–June	Short delayed implantation. Gestation period 40–91 days; 2–6 young born April–May	No delayed implantation. Gestation period 38–45 days; 3–9 young born April–June	Unknown
Remarks	Solitary, 5 extinct species	Crepuscular, species extirpated in many areas	1 extant, 1 extinct species; raised on fur farms	3 extant, 1 extinct species	2 extant species; baculum three-pronged may be a distinct genus

ªM = male; F = female; TL = total length; T = tail. Measurements are in mm.

ica. The weasel lineage started with *M. pliocaenica* in the middle Pliocene, gave rise to *M. praenivalis* in the late Pliocene, and culminated with *M. nivalis,* the weasel (Kurtén 1968). An offshoot of this line is the least weasel, the smallest living carnivore. Although many mammalogists regard it as a dwarf form of *M. nivalis,* others believe that it is a distinct species, *M. rixosa.* In south-central Sweden the 2 forms exist side by side without interbreeding, a condition that has persisted since the early Holocene (about 7,000 years ago) when *M. nivalis* entered the area from the south and *M. rixosa* came in from the east (Kurtén 1972). In America, *M. frenata,* the long-tailed weasel, first appeared in the late Blancan (3.4 mya); it is probably descended from the middle Blancan species, *M. rexroadensis.* Sexual dimorphism is pronounced in weasels, and this plus geographic variation makes identification difficult.

The subgenus *Lutreola* as recently redefined by Youngman (1982) includes 4 species: *M. lutreola, M. siberica, M. lutrolina,* and *M. nudipes.* All show cranial and karyotypic similarities and are semiaquatic. Their fossil history is poorly known: *M. siberica* is known from the Pleistocene and Holocene of Siberia and the Holocene of the Russian plains and Crimea (Vereshchagin and Baryshnikov 1984); *M. lutreola* is only known from the Holocene of Holland (Youngman 1982); nothing is known of the fossil history of the other 2 species. The European mink shows weak geographic variation, and no subspecies are now recognized (Youngman 1982).

The relationship between *M. vison* (the American mink) and *M. lutreola* (the European mink) has been questioned for years, with many workers believing them to be conspecific or at least closely related. Youngman (1982) in his study of *M. lutreola* determined that it is not closely related to *M. vison.* He placed *M. vison* in a separate subgenus *Vison,* along with *M. macrodon,* the extinct sea mink known from Holocene middens along the northeastern coast of the United States. He showed convincingly that *M. vison* is an endemic American mustelid more closely allied to *Martes* than to other species of *Mustela.* Characteristics it shares with *Martes* include the incipient metaconid on M_1, the expanded lingual lobe of M_1, and the well-developed protocone on P^4. Youngman (1982, 33) concluded that "the phenetic, karyological and immunological evidence suggests that the American mink is neither an aquatic polecat nor a near relative of the European mink."

Mustela vison has been found in a few Irvingtonian (early Pleistocene) faunas in the United States and in several late Pleistocene sites, but its remains are never common. It is known from the late Pleistocene Fairbanks and Porcupine River Cave 1 faunas in Alaska, but apparently did not cross into Siberia. The species has been introduced into Europe for fur farming: many animals have escaped and flourished to become a chief competitor and predator of the European mink.

The South American weasels (subgenus *Grammogale*) are regarded as the most primitive of the American weasels (Hall 1951). The braincase is large and flattened anteriorly, with a wide postorbital region and flattened auditory bullae; the tooth row is crowded, with P^2 reduced or absent. The baculum is slender, and the distal end is 3-pronged. Two species have been described, *M. africana* and *M. felipei* (Izor and de la Torre 1978). Relationships to other species of *Mustela* are unclear, and the fossil history is unknown. Some workers believe that *Grammogale* should be given generic rank. About 35 specimens are in collections.

The subgenus *Putorius*, the ferrets or polecats (the names are interchangeable; polecat is usually used for the Old World species), comprises a distinct natural group of medium-sized terrestrial mustelids. Three extant species are currently recognized: *M. putorius*, the European polecat; *M. eversmanni*, the steppe polecat; and *M. nigripes*, the black-footed ferret. The domestic ferret, *M. putorius furo*, has been bred in captivity for more than 2,000 years as a hunter of rodents and rabbits and as a pet.

M. putorius is distinguished from *M. eversmanni* by somewhat smaller size, relatively small canines and carnassial teeth, and more moderate postorbital constriction. *M. eversmanni* and *M. nigripes* show a striking physical resemblance to each other, and their cranial and dental characters are nearly identical (Anderson et al. 1986).

Polecats probably arose in Europe in the Villafranchian (3 to 4 mya) from weasel-like ancestors (Kurtén 1968). The earliest known species, *M. stromeri*, has been found in late Villafranchian to early middle Pleistocene faunas in central Europe. This poorly known species was smaller than *M. putorius;* its mode of life is unknown. It seems probable that it gave rise to *M. putorius* and *M. eversmanni*.

In the middle-Pleistocene European faunas, larger polecats that are identical to the living species made their appearance. Whether the remains belong to *M. putorius* or *M. eversmanni* is difficult to determine because of the fragmentary condition of the specimens and the osteological similarity of the 2 species. Recently Hugueney (1975) identified a complete skull and associated mandible as *M. eversmanni* from La Fage, a late middle-Pleistocene (Riss) fauna in south-central France. This is the earliest known authenticated record of the species in Europe. By the late Pleistocene (Würm), both species were definitely present in Eurasia, with *M. putorius* being more common in European cave faunas and *M. eversmanni* appearing in faunas in Siberia, Crimea, and the Russian plains; it was also a member of Holocene faunas in the Caucasus and central Asia (Vereshchagin and Baryshnikov 1984).

Two species of ferrets are known from the North American Pleistocene. *M. eversmanni* was a member of the late Pleistocene Beringian fauna that ranged across the vast unglaciated steppe that extended from northeastern Siberia to western Alaska. The polecat remains (a partial skull and 2 mandibles) were found near Fairbanks and are characterized by large size, a broad facial region, massive postorbital processes, pronounced postorbital constriction, enlarged canines, and a crowded tooth row. Measurements exceed those of the largest extant subspecies, *M. e. michnoi,* and Anderson (1977) described it as a new subspecies, *M. e. beringiae.* In 1985, additional ferret material, including a well-preserved mummy, was found in caves in the northern Yukon Territory (Youngman, letter to the author, Nov. 1985; telephone conversation, Aug. 1985). The material is under study, and it has not been determined if the remains are those of a steppe polecat or a black-footed ferret. The age of the deposit is about 40,000 years.

The earliest occurrence of *M. nigripes* is uncertain, but the species has probably been present in North America since the Sangamonian (about 100,000 years ago). It has been identified in 22 Pleistocene and early Holocene faunas (Anderson et al. 1986). In 1985, 2 ferret skulls were found in Porcupine Cave, Park County, Colorado, a newly discovered Pleistocene site that has been sealed for at least 10,000 years. Ferrets inhabited South Park until at least the mid-1940s, and this fossil find indicates that they may have made this high mountain valley with its abundant food base their home for thousands of years.

Ferrets arrived in North America in the Sangamonian (perhaps much earlier) from Siberia and spread southward through ice-free corridors to the Great Plains. The rate of range extension for *M. putorius* has been documented in Finland, where between 1880 and 1940 it spread from the Karelian Isthmus north to central Ostrobothnia and west to the Gulf of Bothnia at a rate of 7.5 km per year or 750 km per century (Kalela cited in Kurtén 1957). A similar rate can be postulated for the spread of ferrets during the Pleistocene when conditions were favorable.

No subspecies of *M. nigripes* have been named, and recent studies (Anderson et al. 1986) do not show any taxonomically significant geographic variation between samples. Two or perhaps 3 subspecies of *M. putorius* are recognized based on slight differences in size and coat color. Seventeen subspecies of *M. eversmanni* have been described, 8 of them from Siberia (Stroganov 1962), but whether they are all valid is uncertain since no comparative taxonomic studies have been done. The steppe polecat has by far the largest geographic range: from Hungary to far eastern Asia between 50° and 60° N latitude; Stroganov (1962) notes that it shows more geographic variation than *M. putorius*. A relatively homogeneous environment may have been a factor in the absence of subspeciation in *M. nigripes*.

M. nigripes and *M. eversmanni* are closely related and their possible conspecificity has been noted by several authors (see Youngman 1982 for early references). The 2 species are similar in size and coloration, and analysis of cranial and tooth measurements show only slight differences between them (Anderson et al. 1986). As Anderson (1977,10) noted, "That the two species are closely related cannot be doubted, but until detailed comparative and statistical studies are made on the large collections of *M. eversmanni* in Soviet institutions, the data compared with the information already compiled on *M. nigripes*, and behavioral and chromosomal studies are undertaken on both species, I regard them as distinct." Detailed studies still have not been done on *M. eversmanni*, nor have there been any karyological or immunological studies on the 2 species (but see O'Brien et al., chap. 3, this volume). These studies must be done before the question of conspecificity can be resolved.

References

Anderson, E. 1977. Pleistocene Mustelidae (Mammalia, Carnivora) from Fairbanks, Alaska. *Bull. Mus. Comp. Zool.* 148:1–21.

Anderson, E., S. C. Forrest, T. W. Clark, and L. Richardson. 1986. Paleobiology, biogeography and systematics of the black-footed ferret, *Mustela nigripes* (Audubon and Bachman, 1851). *Great Basin Nat. Memoirs* 8:11–62.

Hall, E. R. 1951. American weasels. *Univ. Kansas Publ. Mus. Nat. Hist.* 4:1–466.

Harrison, J. A. 1981. A review of the extinct wolverine, *Plesiogulo* (Carnivora, Mustelidae) from North America. *Smithsonian Contrib. Paleobiol.* 46:1–27.

Hendey, Q. B. 1978. Late Tertiary Mustelidae (Mammalia, Carnivora) from Langebaanweg, South Africa. *Ann. South African Mus.* 76(10):329–57.

Hugueney, M. 1975. Les Mustélidés (Mammalia, Carnivora) du gisement Pléistocène moyen de la Fage (Correze). *Nouv. Arch. Mus. Hist. Nat. Lyon* 13:29–46.

Izor, R. J., and L. de la Torre. 1978. A new species of weasel (*Mustela*) from the highlands of Columbia, with comments on the evolution and distribution of South American weasels. *J. Mamm.* 59(1):92–102.

Kurtén, B. 1957. Mammal migrations, Cenozoic stratigraphy and the age of Peking Man and the australopithecines. *J. Paleontol.* 31:215–57.

———. 1968. *Pleistocene mammals of Europe.* London: Weidenfeld and Nicolson.

———. 1972. *The ice age.* New York: G. P. Putnam's Sons.

Repenning, C. A. 1976. *Enhydra* and *Enhydriodon* from the Pacific Coast of North America. *J. Res. U.S. Geol. Surv.* 4(3):305–15.

Romer, A. S. 1966. *Vertebrate paleontology*. Chicago and London: Univ. of Chicago Press.

Schmidt-Kittler, N. 1981. Zur Slammesgeschichte der marderverwandten Raubtiergruppen (Musteloidea, Carnivora). *Ecologae geol. Helv.* 74(3):753–801.

Simpson, G. G. 1945. The principals of classification and a classification of mammals. *Bull. Amer. Mus. Nat. Hist.* 85:1–350.

Stroganov, S. U. 1969 [1962]. *Carnivorous mammals of Siberia*. Jerusalem: Israeli Program for Scientific Translation.

Vereshchagin, N. K., and G. F. Baryshnikov. 1984. Quaternary mammalian extinctions in northern Eurasia. In *Quaternary extinctions—a prehistoric revolution,* ed. P. S. Martin and R. G. Klein, 483–516. Tucson: Univ. of Arizona Press.

Youngman, P. M. 1982. Distribution and systematics of the European mink *Mustela lutreola* Linneaus 1761. *Acta Zool. Fennica* 166:1–48.

3

STEPHEN J. O'BRIEN, JANICE S. MARTENSON,
MARY A. EICHELBERGER, E. TOM THORNE,
AND FRANK WRIGHT

Genetic Variation and Molecular Systematics of the Black-Footed Ferret

The black-footed ferret is 1 of 64 species in the family Mustelidae. Taxonomically, it has been placed in the subgenus *Putorius* with 2 other species, *M. putorius* (European polecat, common or domestic ferret) and *M. eversmanni* (steppe or Siberian polecat) (Nowak and Paradiso 1983; Ewer 1973). The black-footed ferret had a wide range in North America from southern Alberta and Saskatchewan to the southwestern United States (Texas-Arizona) as recently as 50 years ago. The number and range of the species have been reduced dramatically over the last century, primarily due to the intentional human eradication of their principal prey base and associate, the prairie dog (*Cynomys* spp). Because of the extreme paucity of black-footed ferrets, the species is listed as endangered under the U.S. Endangered Species Act (Cole, chap. 14, this volume) and is classified as Appendix I by CITES. The black-footed ferret was thought by many to be extinct until it reappeared in 1964 in Mellette County, South Dakota. That population was studied intensively but disappeared in the mid-1970s.

In 1981, a new population of ferrets was discovered near the town of Meeteetse, Wyoming, which is the only known, living relict of the once widespread species (Thorne and Belitsky, chap. 16, this volume). The

We are grateful to Doctors Oliver Ryder, Rodney Mead, Richard Aulerich, and Lyndsay Phillips for providing us with samples of mustelids used in the genetic distance analysis. We are also grateful to Doctors Ulysses S. Seal and Michael Bogan for introducing us to this fascinating biological species. Tissues were collected from exotic animals in full compliance with specific federal fish and wildlife permits (CITES; Endangered and Threatened Species; Captive Bred) issued to Stephen J. O'Brien of the National Cancer Institute, National Institutes of Health, by the U.S. Fish and Wildlife Service of the Department of the Interior.

Meeteetse ferrets reached a peak of 129 animals by 1984. By August 1985, however, the population had declined to approximately 58 animals and continued to drop to an estimated 38 by October of that year. Canine distemper, which originated in the wild, was diagnosed in 2 of 6 ferrets captured in late September and October for captive propagation, and all 6 eventually died. An additional 6 ferrets were captured, but most of the free-ranging population was lost due to the effects of the disease (Thorne and Belitsky, chap. 16, this volume). During the summer and fall of 1986, 11 additional ferrets were captured because it was believed that the population was facing imminent extinction in the wild (Thorne and Belitsky, chap. 16, this volume). Currently, 18 black-footed ferrets are in captivity. The fate of the free-ranging population is unknown, but it cannot be considered viable.

The low numbers of remaining black-footed ferrets pose a severe threat to species survival for at least two reasons. The first is the potential for a demographic crash. Within small populations, new generations may not be produced simply because of chance effects (accidents, altered sex ratio, damage to prey base, diseases). The canine distemper epizootic is a graphic example of these effects on the black-footed ferret. A second concern is that in order to reproduce, small populations must inbreed by necessity, producing "founder effects" with associated genetic consequences. These include inbreeding depression because of homozygous expression of deleterious recessive genes which can affect reproduction, fecundity, development, and survivorship (Ralls et al. 1979; Ralls and Ballou 1982; O'Brien et al. 1985, 1986; O'Brien and Knight 1987). In addition, genetic homogenization by forced inbreeding removes endemic genetic variation in genetic loci encoding immune defense mechanisms. This translates into a more widespread population (or species) sensitivity to viral, bacterial, or parasitic agents in the environment, which have by molecular adaptation developed virulence against an individual's genotype. Such severe consequences of genetic uniformity have been observed with another threatened species, the cheetah (O'Brien et al. 1985, 1986; O'Brien and Evermann 1988).

Because of the troublesome natural history of the black-footed ferret, we sought to examine the genetic status of the only known population. An estimate of the extent of genetic variability in this population was derived using a survey of 46 gene-enzyme systems previously used in our laboratory to study genetic variability in other carnivore species. In addition, we have compared the electrophoretic mobility of homologous enzyme systems of four species of *Mustela* (*M. nigripes*, *M. putorius*, *M. eversmanni*, and *M. vison*) in an attempt to resolve the phylogenetic relationship among these species. We have used these results to estimate the time elapsed since the black-footed ferret and its closest relative, the steppe polecat (*M. eversmanni*), have shared a common ancestor.

Materials and Methods

Heparinized blood and skin biopsies were obtained from the animals listed in Table 3.1. Blood was separated into erythrocytes, leukocytes, and plasma. Skin biopsies were digested with trypsin and collagenase and used to establish primary fibroblast cultures (Modi et al. 1987). Isozyme extracts of blood components and tissue culture cells were prepared by sonication. For the population survey of *M. nigripes*, isozyme extracts of 3 tissues (liver, kidney, and spleen) from 6 ferrets that died of canine distemper were used, in addition to blood products from 6 live, captive animals. Electrophoretic procedures for separation and resolution of isozyme systems have been described previously (Newman et al. 1985; O'Brien 1980; Harris and Hopkinson 1976). The genetic distance values were computed based on 41 isozyme systems resolved in 4 tissues: erythrocytes, leukocytes, plasma, and cultured fibroblasts using the formulae of Nei (1972, 1978). Evolutionary trees were constructed using the UPGMA algorithm (Sneath and Sokal 1973) of the BIOSYS computer program (Swofford 1981), the Wagner distance method (Farris 1972; Swofford and Selander 1981), and the Fitch-Margoliash algorithm (Fitch and Margoliash 1967) of the PHYLIP computer program (Felsenstein 1984).

In an electrophoretic survey of 12 individual black-footed ferrets from the Meeteetse population, we found one polymorphic locus (out of 46 tested). Thus, the percentage of polymorphism (*P*) is 2%, and the average heterozygosity (*H*) equals 0.008 (Table 3.2). These values are lower than estimates for each of 10 other carnivore species tested with the same enzyme systems (Table 3.2), with the single exception of the South African cheetah (O'Brien et al. 1983, 1985, 1987c). Two of these feline species, ocelot and margay, actually had fewer animals in the survey but abundant genetic variation (Newman et al. 1985.). Two recently examined species of procyonids (raccoons and kinkajous) had the highest levels of genetic variation (*P* = 33–39%) in carnivores (Forman 1985).

Early studies by Simonsen (1982) and Allendorf (1979) failed to observe electrophoretic variation in certain carnivore species, including several mustelids and the black bear (Table 3.2). These studies examined 21 and 13 loci, respectively, and for this reason are not strictly comparable to our results. This is especially true because certain isozyme

Table 3.1. Animals Used in Population and Genetic Distance Studies

Latin Name	Common Name	Tissue Used	No.	Source
Mustela eversmanni	Siberian polecat	Blood, skin	2	Minnesota Zoological Gardens[1]
Mustela eversmanni	Siberian polecat	Blood, skin	1	Wyoming Game & Fish Dept.[2]
Mustela putorius	European polecat	Blood	2	Minnesota Zoological Gardens[1]
Mustela putorius	European polecat	Blood, skin	2	Wyoming Game & Fish Dept.[2]
Mustela putorius	European polecat	Organs	2	Wyoming Game & Fish Dept.[2]
Mustela nigripes	Black-footed ferret	Blood, skin	6	Wyoming Game & Fish Dept.[2]
Mustela nigripes	Black-footed ferret	Organs	6	Wyoming Game & Fish Dept.[2]
Mustela nigripes	Black-footed ferret	Skin cell culture	1	San Diego Zoo[3]
Mustela vison	Mink	Blood	6	Michigan State University[4]
Mustela vison	Mink	Skin cell culture	1	American Type Culture Collection
Mustela vison	Mink	Lung cell culture	1	American Type Culture Collection
Spilogale putorius	Spotted skunk	Blood, skin	4	University of Idaho[5]
Ictonyx striatus	African striped skunk	Blood, skin	1	National Zoological Park[6]
Ursus americanus	American black bear	Blood, skin	2	Wild-caught, Pocono Mts., Penn.[7]
Ursus americanus	American black bear	Blood, skin	2	Wild-caught, Grand Rapids, Minn.[8]
Mephitis mephitis	Striped skunk	Kidney	1	Wild-caught, Frederick, Md.[9]

[1]Dr. Frank Wright, [2]Dr. Tom Thorne, [3]Dr. Oliver Ryder, [4]Dr. Richard Aulerich, [5]Dr. Rodney Mead, [6]Dr. Lyndsay Phillips, [7]Dr. Olaf Oftadal, [8]Dr. Dave Garshelis, [9]Dr. Stephen O'Brien

Table 3.2. Comparison of Electrophoretic Estimates of Biochemical Variation among Carnivore Species

Species	Common Name	Populations (no.)	Individuals (no.)	Loci (no.)	Loci Estimated Polymorphic (%)	Average Heterozygosity	References
Mustelidae							
Mustela nigripes	Black-footed ferret	1	12	46	2	0.008	This study
Mustela erminea	Stoat	1	39	21	0	0.0	Simonsen 1982
Mustela nivalis	Weasel	1	13	21	0	0.0	Simonsen 1982
Mustela putorius	Polecat	1	24	21	0	0.0	Simonsen 1982
Martes martes	Pine martin	1	2	21	0	0.0	Simonsen 1982
Martes foina	Beech martin	1	121	21	0	0.0	Simonsen 1982
Meles meles	Badger	1	5	21	0	0.0	Simonsen 1982
Canidae							
Vulpes vulpes	Red fox	1	282	21	0	0.0	Simonsen 1982
Ursidae							
Ursus americanus	Black bear	6	233	35	18	0.011	Manlove et al. 1977
Ursus maritimus	Polar bear	2	52	13	0	0.0	Allendorf et al. 1979
Procyonidae							
Procyon lotor	Raccoon	23	526	49	33	0.035	Beck & Kennedy 1980; Dew & Kennedy 1980; Forman 1985
Potos flavus	Kinkajou	1	27	33	39	0.132	Forman 1985
Felidae							
Felis catus	Domestic cat	1	56	61	21	0.082	O'Brien 1980
Leopardus pardalis	Ocelot	3	6	48	21	0.072	Newman et al. 1985
Leopardus weidi	Margay	3	11	50	16	0.047	Newman et al. 1985
Caracal caracal	Caracal	2	16	50	10	0.029	Newman et al. 1985
Leptailurus	Serval	3	16	49	12	0.033	Newman et al. 1985
Panthera pardus	Leopard	5	18	50	8	0.029	Newman et al. 1985
Panthera leo	Lion	6	42	40	18	0.050	Newman et al. 1985
Panthera tigris	Tiger	5	40	50	10	0.035	Newman et al. 1985
Neofelis nebulosa	Clouded leopard	5	25	49	6	0.023	Newman et al. 1985
Acinonyx j. jubatus	Cheetah, S. African	7	55	52	0	0.0	O'Brien et al. 1985
Acinonyx j. raineyi	Cheetah, E. African	2	30	49	4	0.014	O'Brien et al. 1987c

loci that are sampled in electrophoretic surveys tend to be polymorphic in mammals, while other loci are not (O'Brien et al. 1980; Wayne et al. 1986). Because of the lower number of typed loci in these early studies, several of the more polymorphic loci were not examined.

For these reasons, we interpret our limited survey of the Meeteetse black-footed ferret population as showing a low level of genetic variation relative to other carnivore species, including small carnivores, similarly studied. For example, the amount of variation is comparable to the levels observed in the east African cheetahs (Table 3.2; O'Brien et al. 1987c). These results confirm the theoretical expectation of low variability in natural populations that have experienced severe founder effects in their recent history. The reproductive, genetic, and epi-

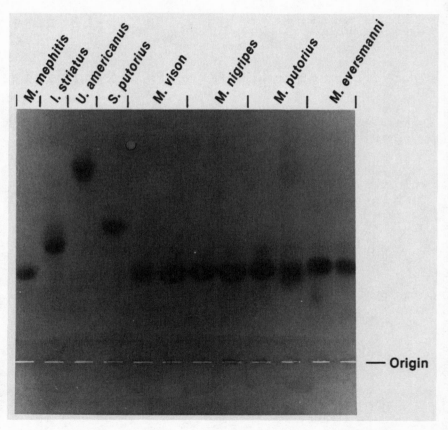

Figure 3.1. Electropherogram of a starch gel loaded with extracts of indicated species and developed histochemically for 6-phosphogluconate dehydrogenase.

demiological consequences of this situation are not immediately clear. The primary threat to the population today is demographic, but genetic status can influence the recovery from these events.

The evolutionary relationship between the black-footed ferret and several other mustelids, including *M. eversmanni, M. putorius,* and *M. vison,* was estimated by computing the isozyme-genetic distance. In this procedure, the electrophoretic mobility of homologous enzymes from different species is compared, and the extent of mutational accumulation in the protein primary structure is estimated (Fig. 3.1). The genetic distance, *D,* is defined as the average number of gene differences per locus between individuals in 2 test populations (Nei 1972, 1978). Within the limits of certain assumptions relating to the electrophoretic resolution and relative rates of amino acid substitutions, the genetic distance values increase proportionately with the time elapsed since the compared populations became reproductively isolated.

The electrophoretic phenotypes of 41 informative isozyme systems for 8 tested species are presented in Table 3.3. These data were used to compute a matrix of unbiased genetic distance, *D,* and minimum unbiased genetic distance between each species (Table 3.4). Several phylogenetic algorithms were employed to construct evolutionary trees relating the species. The results of these analyses produced phylogenetic trees that were topologically equivalent but different in limb length. Two derived phylogenetic trees are presented in Figure 3.2.

The molecular phylogeny of the mustelid species indicates a rather ancient split, 20 to 30 million years ago, between skunks and the genus *Mustela.* More recently, the mink (*M. vison*) diverged from the *Putorius* subgenus. The 3 ferret species apparently diverged during the early Pleistocene or late Pliocene, 1 to 3 million years ago.

The molecular data affirm that the black-footed ferret's closest relative is *M. eversmanni,* as several authors had suggested on the basis of morphology (Rempe 1965; Anderson 1977; Anderson et al. 1986; Anderson, chap. 2, this volume). In fact, certain authors had suggested that the 2 species are so similar as to raise the possibility that the 2 ferrets represent conspecific members of the same species or zoogeographic subspecies (Anderson 1977; Anderson et al. 1986).

The present data can comment on these conclusions by comparing the genetic distance between *M. nigripes* and *M. eversmanni* (*D* = 0.081, Table 3.4) to similar values obtained for other carnivore groups. The *M. nigripes-M. eversmanni* distance is greater than the genetic distances measured in our laboratory with the same loci between 3 subspecies of tiger (average *D* = 0.007; range = 0.003–0.01), between Asian versus African lion subspecies (mean *D* = 0.014; range = 0.007–0.033), between East

Table 3.3. Relative Electrophoretic Mobility of Isozymes among Mustelid Species

Isozyme Marker[a]	Species							
	MNI[b]	MEV	MPU	MVI	SPU	UAM	IST	MME
ACP1	A[c]	A	A	A	A	B	A	—[d]
ACP3	A	A	A	A	A	B	A	A
ADA	B	B	A	C	D	E	D	D
AK1	B	A	A	B	C	C	C	C
ALB	A	A	A	A	—	—	B	—
APRT	A	A	B	B	C	—	—	—
CA2	A	A	A	A	A	B	A	—
CPKB	A	A	A	B	B	B	B	B
DIA1	A	A	A	E	B	C	D	C
ESD	A	A	A	A	B	C	A	CB
ES2	B	A	A	A	—	A	—	—
GOT1	A	A	A	A	—	B	—	—
GOT2	A	A	A	A	A	A	A	A
G6PD	A	A	A	C	D	A	AB	D
GPI	A	A	A	B	C	D	E	F
GSR	A	A	A	A	A	B	A	A
GPT	A	A	B	—	—	—	—	—
GLO	A	A	A	A	A	B	A	A
HBB	B	A	A	A	B	C	D	—
IDH1	A	A	A	A	—	B	—	—
ITPA	A	A	B	B	B	C	D	E
LDHA	A	A	A	A	B	C	A	B
LDHB	A	A	A	A	A	A	A	A
MDH1	A	A	A	A	A	A	A	A
MDH2	A	A	B	A	A	C	A	A
ME1	A	AB	B	B	C	F	D	E
MPI	A	A	A	B	C	D	E	F
NP	A	A	A	A	C	D	B	E
PEPB	A	A	A	AB	C	D	D	D
PEPC	A	A	A	A	—	B	C	C
PEPD	A	A	A	A	B	C	D	B
PGD	A	A	A	A	B	C	D	A
PGAM	A	A	A	AB	A	A	A	A
PGM1	A	A	A	A	B	C	D	B
PGM2	A	A	A	A	BC	A	A	C
PGM3	A	A	A	B	C	D	C	D
PK	A	A	A	A	B	B	B	C
PP	A	A	A	A	C	D	D	—
SOD1	A	A	A	D	B	C	E	F
TPI	A	A	A	A	A	A	A	A
TF	A	A	A	A	C	—	B	—

[a]Names of enzymes follow human nomenclature (Shows et al. 1985).
[b]Abbreviations for column headings are: MNI = *Mustela nigripes*; MEV = *Mustela eversmanni*; MPU = *Mustela putorius*; MVI = *Mustela vison*; SPU = *Spilogale putorius*; UAM = *Ursus americanus*; IST = *Ictonyx striatus*; and MME = *Mephitis mephitis*.
[c]Letters represent electrophoretic types for the designated enzyme. Single letter indicates that this species was monomorphic for a single allelic form; double letters indicate that two allelic forms were observed.
[d](—) indicates no result.

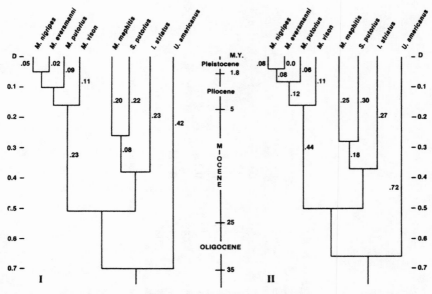

Figure 3.2. Two evolutionary trees constructed using the isozyme genetic distance data in Table 3.4. Tree number I was derived from Nei's unbiased minimum distance (1978) and the UPGMA cluster (Sneath and Sokal 1973) algorithm. This algorithm derives a tree with contemporaneous extant species. The limb lengths were derived from the Wagner distance algorithm from the same program that constructs an unrooted topology using Nei's unbiased genetic distance (1978). Tree number II was drawn by the PHYLIP computer program of J. Felsenstein (University of Washington) using the Fitch-Margoliash algorithm (1967), also assuming contemporary living species or tips. The limb length of tree II was derived from the same algorithm (Fitch) in the absence of this assumption. The tree is calibrated with 2 dates. The first point is based on the estimated appearance of the great cats, *Panthera*, approximately 1.8 million years ago (Neff 1983; Savage and Russell 1983). The mean distance of 10 pairwise comparisons of *Panthera* species was 0.101, so this *D* value was set as the interface of the Pleistocene and Pliocene. The second date, 35–40 million years ago, is the time of divergence of carnivore families, specifically the Ursidae and Mustelidae. These 2 dates do not provide a linear relationship; however, they represent fairly accurate limits to the time scale of the presented phylogenies.

African versus South African cheetahs ($D = 0.004$), between 3 lion tamarin subspecies (mean $D = 0.016$; range $= 0.007–0.03$), and between human racial groups (mean $D = 0.022$; range $= 0.01–0.029$) (O'Brien et al. 1987a, 1987b, 1987c; Forman et al. 1986; Nei and Roychoudhury 1974). This result would suggest that *M. nigripes* and *M. eversmanni* have been reproductively isolated for a period 5 to 10 times longer than the other subspecies.

The average genetic distance between the 5 different species of great cats (*Panthera*) was estimated previously as 0.103, with the range equal to 0.03–0.23 (O'Brien et al. 1987a). The distance between the 2 ferret

Table 3.4.   Genetic Distance between Mustelid Species Based on Isoenzyme Analysis

	M. nigripes	M. eversmanni	M. putorius	M. vison	M. mephitis	S. putorius	I. striatus	U. americanus
M. nigripes		0.077	0.220	0.317	0.656	0.657	0.588	0.811
M. eversmanni	0.081		0.126	0.271	0.647	0.678	0.580	0.776
M. putorius	0.248	0.136		0.242	0.689	0.686	0.616	0.784
M. vison	0.385	0.322	0.279		0.614	0.620	0.559	0.775
M. mephitis	1.097	1.079	1.193	0.984		0.420	0.513	0.639
S. putorius	1.070	1.149	1.157	0.982	0.553		0.520	0.758
I. striatus	0.900	0.892	0.971	0.841	0.742	0.744		0.653
U. americanus	1.665	1.523	1.531	1.523	1.039	1.417	1.075	

Note: Above diagonal, Nei-unbiased minimum genetic distance; below diagonal, Nei-unbiased genetic distance (D) corrected for low sample size (Nei, 1978). Matrix was computed using BIOSYS Program (Swofford 1981).

types falls in the lower end of this range and would be consistent with a separation at about the same time that the great cats diverged. The molecular data interpreted by comparative analysis with other carnivore species would be consistent with a distinct species designation of *M. nigripes* and *M. eversmanni* and a separation time of 0.5 to 2 million years ago.

References

Allendorf, F. W., et al. 1979. Electrophoretic variation in large mammals. I. The polar bear, *Thalarctos maritimus. Hereditas* 91:19–22.

Anderson, E. 1977. Pleistocene Mustelidae (Mammalia, Carnivora) from Fairbanks, Alaska. *Bull. Mus. Comp. Zool.* 148:1–21.

Anderson, E., et al. 1986. Paleobiology, biogeography and systematics of the black-footed ferret, *Mustela nigripes* (Audubon and Bachman). *Great Basin Nat. Memoirs* 8:11–62.

Beck, M. L., and M. L. Kennedy. 1980. Biochemical genetics of the raccoon, *Procyon lotor. Genetica* 54:127–32.

Dew, R. D., and M. L. Kennedy. 1980. Genetic variation in raccoons, *Procyon lotor. J. Mammal.* 61:697–702.

Ewer, R. F. 1973. *The carnivores.* Ithaca, New York: Cornell Univ. Press.

Farris, J. S. 1972. Estimating phylogenetic trees from distance matrices. *Am. Nat.* 106:645–88.

Felsenstein, J. 1984. Distance methods for inferring phylogenies: A justification. *Evolution* 38:16–24.

Fitch, W. M., and E. Margoliash. 1967. Construction of phylogenetic trees. *Science* 155:279–84.

Forman, L. 1985. Genetic variation in two procyonids: Phylogenetic, ecological, social correlates. Ph.D. diss., New York Univ.

Forman, L., et al. 1986. Genetic variation within and among lion tamarins. *Am. J. Phys. Anthropol.* 71:1–11.

Harris, H., and D. A. Hopkinson. 1976. *Handbook of enzyme electrophoresis in human genetics.* Amsterdam: North Holland Publishing Co.

Manlove, M. N., et al. 1977. Biochemical variation in the black bear. *The Bear Biology Association Conference Series.* 3:37–41.

Modi, W. S., et al. 1987. Cytogenetic methodologies for gene mapping and chromosomal analyses in mammalian cell culture systems. *Gene Anal. Tech.* 4:75–85.

Neff, N. 1983. *The big cats: The paintings of Guy Coheleach.* New York: Abrams.

Nei, M. 1972. Genetic distance between populations. *Am. Nat.* 106:283–92.

Nei, M. 1978. Estimation of average heterozygosity and genetic distance from a small number of individuals. *Genetics* 89:583–90.

Nei, M., and A. K. Roychoudhury. 1974. Genetic variation within and between the three major races of man, Caucasoids, Negroids and Mongoloids. *Am. J. Hum. Genet.* 26:421–43.

Newman, A., et al. 1985. Biochemical genetic variation in eight endangered feline species. *J. Mammal.* 66:256–67.

Nowak, R. M., and J. L. Paradiso. 1983. *Walker's mammals of the world.* 4th ed., 1061–94. Baltimore and London: The Johns Hopkins Univ. Press.

O'Brien, S. J. 1980. The extent and character of biochemical genetic variation in the domestic cat (*Felis catus*). *J. Hered.* 71:2–8.

O'Brien, S. J., and J. F. Evermann. 1988. The interface of epidemiology and genetic diversity in free-ranging animal populations. *Trends Ecol. Evolution.* 3:254–59.

O'Brien, S. J., and J. A. Knight. 1987. The giant panda's future. *Nature* 325:758–59.

O'Brien, S. J., et al. 1980. Correlative genetic variation in natural populations of cats, mice and men. *Nature* 288:580–83.

O'Brien, S. J., et al. 1983. The cheetah is depauperate in biochemical genetic variation. *Science* 221:459–62.

O'Brien, S. J., et al. 1985. Genetic basis for species vulnerability in the cheetah. *Science* 227:1428–34.

O'Brien, S. J., et al. 1986. The cheetah in genetic peril. *Sci. Am.* 254:84–92.

O'Brien, S. J., et al. 1987a. Setting the molecular clock in Felidae: The great cats, *Panthera.* In *Tigers of the world,* ed. R. L. Tilson and U. S. Seal, 10–27. Park Ridge, N.J.: Noyes.

O'Brien, S. J., et al. 1987b. Biochemical genetic variation in zoogeographic isolates of African and Asiatic lions. *Natl. Geo. Res.* 3:114–24.

O'Brien, S. J., et al. 1987c. East African cheetahs: Evidence for two population bottlenecks? *Proc. Natl. Acad. Sci. USA* 84:508–11.

Ralls, K., and J. Ballou. 1982. Effects of inbreeding on juvenile mortality in some small mammal species. *Lab. Anim.* 16:159–66.

Ralls, K., et al. 1979. Inbreeding and juvenile mortality in small populations of ungulates. *Science* 206:1101–03.

von Rempe, Udo. 1965. Lassen sich bei Saugetieren Introgressionen mit multivarianten Verfahren nachweisen? *Zeitschrift Zool. System. Evolut.* 3:(S) 388–412.

Savage, D. E., and D. E. Russell. 1983. *Mammalian paleofaunas of the world.* Reading, Mass.: Addison-Wesley.

Shows, T. B., et al. 1987. Guidelines for human gene nomenclature: An international system for human gene nomenclature (ISGN 1987). *Cytogen. Cell Genet.* 46:11–28.

Simonsen, V. 1982. Electrophoretic variation in large mammals. II. The red fox, *Vulpes vulpes,* the stoat, *Mustela erminea,* the weasel, *Mustela nivalis,* the polecat, *Mustela putorius,* the pine marten, *Martes martes,* the beech marten, *Martes foina,* and the badger, *Meles meles. Hereditas* 96:299–305.

Sneath, P. H. A., and R. R. Sokal. 1973. *Numerical taxonomy: The principles and practice of numerical classification.* San Francisco: W. H. Freeman and Co.

Swofford, D. L. 1981. On the utility of the distance Wagner procedure. In

Advances in cladistics, ed. V. A. Funk and D. R. Brooks, 25–43. New York: New York Botanical Garden.

Swofford, D. L., and R. B. Selander. 1981. BIOSYS-1; A Fortran program for the comprehensive analysis of electrophoretic data in population genetics and systematics. *J. Hered.* 72:281–83.

Wayne, R. K., et al. 1986. Genetic monitors of captive zoological populations: Morphological and electrophoretic assays. *Zoo Biol.* 5:215–32.

PART TWO

Population Biology

4

PETER F. BRUSSARD AND MICHAEL E. GILPIN

Demographic and Genetic Problems of Small Populations

In general, rare, threatened, or endangered species have populations that either are sparsely distributed over a large area or exist as small groups in one or a few particular spots; both cases involve small numbers of animals. Population biologists have only recently become interested in the problems of small or fragmented populations, primarily in response to the needs of various governmental agencies charged with preserving biological diversity. For example, the requirement of managing national forest habitats to maintain "viable populations of wildlife and fish" (Salwasser et al. 1984) led to a series of workshops sponsored by the U.S. Forest Service that explored the concept of minimum viable population size. These workshops stimulated considerable theoretical work on this problem. (A book on this topic edited by M. E. Soule has recently been published.) In this chapter, we will summarize our current understanding of the demographic and genetic problems encountered by small populations, using the black-footed ferret as an example whenever possible.

The most important lesson concerning the biology of small populations is that stochastic processes play a critical role in their survival. These processes separate naturally into 3 categories: (1) populational or demographic uncertainty, (2) environmental uncertainty, and (3) genetic uncertainty. Uncertain demographic factors include random variations in sex ratio, age of first reproduction, number of offspring, distribution of offspring over the lifetime of an individual, and age at death. Uncertain environmental factors range from major catastrophes (fires, earthquakes, major storms, pandemic diseases, and so on) to mildly unpredictable environmental variations such as year-to-year changes in weather patterns. Genetic uncertainty includes inbreeding (the mating of related individuals) and loss of variation through genetic drift.

In the most general terms, the short- and mid-term survival potential of a population is determined primarily by its resilience and fitness. Resilience refers to the ability of a population to recover from declines resulting from random variation in normal birth and death events and from environmental perturbations. Resilience is determined by age structure, reproductive potential, social system, and generation time. The fitness of a population depends on its having the appropriate set of genes to cope with its environment; this in turn is largely determined by the presence of sufficient genetic variation (heterozygosity) to maintain normal fecundity and viability under the prevailing ecological circumstances. Beyond this, the long-term survival potential of a population is related to its adaptability, or its ability to evolve. Adaptability depends upon the maintenance of sufficient genetic variation to permit adjustment to environmental change through the process of natural selection.

Species populations normally have considerable capacity to resist threats to their survival through various responses—that is, they have adequate resilience, fitness, and adaptability. Very rare species, however, have much less capacity to resist those threats because of their increased susceptibility to the deleterious effects of stochastic events.

Population Dynamics and Persistence

Demographers usually model population growth by variations of a differential (or difference) equation for change of population size. The first model, continuous exponential growth,

$$dN/dt = rN \tag{4.1}$$

where N is the number of individuals in the population and r is the instantaneous rate of increase, is a fair estimator of population increase in an unlimited environment. Many populations have been observed to grow exponentially for a time. Eventually, however, something puts the brakes on population growth and numbers decline, sometimes precipitously.

A second model of population growth is referred to as the logistic equation. Under this model a population grows exponentially until it reaches a point where the rate of growth begins to decline as resources become increasingly limiting; that is, the growth rate shows density dependence. Eventually the population stabilizes at a point called the carrying capacity (K), defined as the average number of individuals that a given area can support. This model is

$$dN/dt = rN - rN^2/K = rN[(K - N)/K] \qquad (4.2)$$

Even though the logistic equation has often been used to model wild-life populations, it is not universally applicable since relatively few populations have been observed to rise to a certain equilibrium level and then remain constant over time. Rather, they fluctuate, often severely. A major reason for this is that the values of r and K change, both deterministically and stochastically.

Fluctuations in r are at the heart of the population persistence problem. Any factor contributing to a decrease in its growth rate is a potential threat to the continued existence of a population; however, most natural populations continually face such changes without going extinct. This is because some of these factors may be density-dependent—that is, they operate less intensively as population size decreases. Or, the operation of some negative factors may be variable in time or space so that they either diminish before exterminating a population or affect only a segment of it at a time. Thus, most fluctuations in r usually have little long-term effect on the size and persistence probabilities of large, well-distributed populations. As populations become smaller, however, the risk of extinction increases dramatically.

MacArthur and Wilson (1967) and MacArthur (1972) modeled the extinction process as a function of 3 parameters: r, K, and the per-capita birth rate, b, where the per-capita death rate is $r - b$. They predicted that populations with high growth rates, high per-capita birth rates, and large K's (in the thousands) had expected times-to-extinction that were very large. However, for populations with lower r's and b's, extinction was likely to be fairly rapid below a threshold value of K that was in the tens to low hundreds, depending on the values of the former parameters. A model by Richter-Dyn and Goel (1972) also predicted a threshold effect for population persistence of approximately the same order of magnitude. (It is important to note that none of these models considers the effects of age structure on the population; thus, for these models we have to interpret the predicted values as the numbers of breeding individuals, not total population size).

Groves and Clark (1986) used the MacArthur and Wilson (1967) model to predict the time-to-extinction for the Meeteetse black-footed ferret population. Assuming a K of 40 adults, an even sex ratio, and a ratio of death rate to birth rate of 0.8, they predicted that this population should persist for 1,000 years. This number is unrealistic for reasons explored below.

Population Uncertainty and Persistence

The MacArthur and Wilson model does not incorporate any variability in growth rates that might result from random variation in birth and death rates for the entire population; hence, its applicability to natural populations is limited. May (1973) continued the analysis of these factors, now referred to as "individual demographic stochasticity" or V_i, into a second-generation model and concluded that random fluctuations in population size should be roughly proportional to $1/\sqrt{K}$. This implies that V_i will result in a population of about 36 individuals varying between 30 and 42 (36 $\pm$ $\sqrt{K}$, or 6).

More recent models by Leigh (1981) and Goodman (1987a, 1987b) tend to support the idea that V_i is an important factor only for the survival of very small populations and that the importance of V_i declines rapidly with increases in population size. This is because individual variations in survivorship or reproduction tend to be compensated for by the contributions of more and more individuals. As populations achieve relatively modest numbers, predicted times-to-extinction on the basis of V_i alone become quite large.

It would seem, then, that the effects of individual demographic stochasticity would be unlikely to cause the extinction of a population the size of the one at Meeteetse, provided that there was an even sex ratio. However, Forrest et al. (pers. comm.) reported that the adult sex ratio (males to females) in this population was closer to 1 : 2.2 than to unity, suggesting that V_i could pose some problems for population persistence.

Environmental Uncertainty and Persistence

Studies by Leigh (1981) and Goodman (1987a, 1987b) make two very important points. The first is that the expected survival time of a population depends critically on the variability of its growth rate. Second, this variability actually consists of 2 components: individual demographic stochasticity (V_i) and environmental stochasticity or V_e. Environmental stochasticity poses problems for population persistence that can be severe since it is essentially independent of population size. Examples of this variation are population-wide changes in the probabilities of death or reproduction related to the vagaries of climate, disease, competition, predation, or resource availability. The epizootic of canine distemper at Meeteetse during 1985 provides an excellent, and tragic, example of how environmental stochasticity can drastically affect small populations.

Goodman (1987a) has shown that with realistic estimates of the affects of V_e factored in, times-to-extinction become much shorter and much less responsive to higher values of K than those predicted by models that either ignore any variance in growth rate or only consider the effects of V_i. Furthermore, environmental stochasticity results in times-to-extinction that increase only gradually with population size, rather than exhibiting threshold effects. If environmental variability is high, very large K's are required to achieve reasonably long times-to-extinction. Even an increasing population can have a relatively short predicted time to extinction if it has a low rate of increase and a relatively high variance in this rate. Likewise, density dependence in mean growth rate makes little difference in estimated time-to-extinction compared to a calculation with density independent of growth.

Goodman's model can be used to calculate a first-order estimation of time-to-extinction for any population for which the following data are available: (1) estimates of total census size for at least 4 years if an even sex ratio can be assumed, and (2) an estimate of the average per-capita birth rate. Alternatively, if sex ratios are skewed, corresponding estimates of the total number of females in the population and the number of female offspring per female can be used. Unfortunately, the appropriate data are not available from the Meeteetse population prior to its crash to estimate time-to-extinction in that population.

Even though a population consisting of 30 to 40 breeding individuals with an even sex ratio is normally at little risk of extinction from the effects of individual demographic stochasticity, the skewed adult sex ratio observed at Meeteetse may suggest that V_i could be an important factor in the persistence of small populations of black-footed ferrets. A more important consideration, however, is that populations of this size obviously cannot withstand the effects of environmental uncertainty. This was clearly indicated by the collapse of the Meeteetse population after the 1985 distemper epizootic. There is, moreover, a stronger demonstration that the average lifetime of a black-footed ferret population is short and that the collapse at Meeteetse should not have been surprising.

Over the last half century a very large number of black-footed ferret colonies have gone extinct. Some extinctions have resulted from direct human disturbance, but others have not. Assume that there were C such colonies and that this number has been reduced to one over the last Y years. Assume further that each colony behaved independently of all others. It is possible to calculate a "colony half-life," that is, the time it would take for half of the original colonies to have gone extinct, using a simple model of exponential decay. Such calculations are shown in Table 4.1. Over a large span of both C and Y values, colony half-lives range from about 1 to 10 years. Even under the most favorable assumptions (100 initial colonies and 90 years), the average colony half-life is only about 13 years. Thus, even if the Meeteetse population had not collapsed in 1985, we can predict that its short-term (years to decades) persistence would be extremely unlikely and that the population has always been at serious risk of extinction.

Local extinctions and recolonizations are probably typical of the dynamics of the black-footed ferret, and a self-sustaining population of this species in presettlement days was almost certainly a "metapopulation"—a system of small local units, each of which had its own stochastic dynamics. This suggests that multiple reserves, spaced far enough apart so that the environmental variation in them is at least partially independent, should be established for the black-footed ferret.

We have performed a preliminary investigation of the consequences of such a metapopulation structure using a graphics-oriented micro-

Table 4.1. Theoretical Half-Lives for Black-Footed Ferret Colonies

Original Colonies (C) (no.)	Years (Y) to Single Colony[a]	Colony Half-Life[b] (yrs.)
	10	1.51
	30	4.52
100	50	7.53
	70	10.54
	90	13.55
	10	1.00
	30	3.01
1,000	50	5.02
	70	7.02
	90	9.03
	10	0.75
	30	2.26
10,000	50	3.76
	70	5.27
	90	6.77

[a]Number of years over which original colonies were reduced to 1.
[b]Number of years in which half of the original colonies will go extinct.

computer model of metapopulation dynamics. This program allows one to configure the size and separation of the patches on which a meta-population exists; this thereby controls the rates of patch extinction and recolonization. Figure 4.1 shows the graphic output from 1 run of the model. In this run 12 patches were used, and the size of the patches was adjusted to give mean time to patch (local population) extinction of about 10 years, which according to the argument above should be rea-sonable for the black-footed ferret. The patches are shown at the right of the figure in the area labeled "Metapopulation Arena." During the run the patches fluctuate between solid circles, denoting an extant pop-ulation, and open circles, denoting an extinct population. At the left, the history of the run is recorded, by showing both the fraction of occupied patches and the continuity of occupancy of individual patches, which reveals turnover. The metapopulation has gone extinct at generation (year) 128. In this run, the patches were moved very close together to raise the probability of colonization to a high level, but even this arrange-ment was insufficient to sustain the metapopulation.

Our simulations indicate that managers will have to do one of three things to provide for the long-term persistence of a metapopulation of

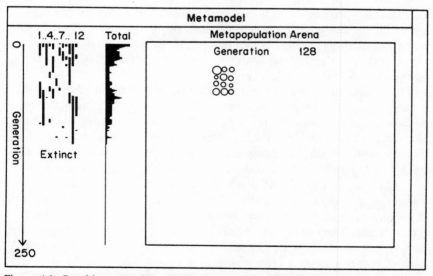

Figure 4.1. Graphic output from 1 run of a model of the metapopulation dynamics of black-footed ferrets based upon 12 patches and mean time to patch extinction of 10 years. The history of the run (shown on the left) indicates the continuity of occupancy of indi-vidual patches and the fraction of occupied patches each generation. Extinction occurred at generation 128.

black-footed ferrets: (1) use a large number of patches located reasonably close together; (2) place the patches very close together so that between-patch colonization rates are on the order of one per 10 years; or (3) manage to recolonize extinct patches. The last option is likely to be the least expensive in the intermediate future.

Genetic Uncertainty and Population Persistence

In addition to random populational and environmental events, genetic uncertainty can threaten population viability. Genetic uncertainty refers to random changes in the genetic makeup of a population that have deleterious effects on the ability of individuals to survive and reproduce, as well as on the capacity of populations to respond adaptively to changes in their environments. The two major genetic factors causing deleterious effects are inbreeding depression and genetic drift. In general, to maintain population fitness and adaptability, a population must be large enough to prevent intense inbreeding and serious reduction in genetic variation by random drift, but these effects must be viewed in the context of the history of the species.

In natural populations the number of individuals actually involved in contributing progeny, and therefore genes, to subsequent generations is generally only a small fraction of the total population size. For example, some populations of checkerspot butterflies, *Euphydryas* spp., which contain several hundred individuals (N), have genetically effective sizes (N_e) in the low tens (Brussard, unpubl. data).

The genetically effective size of a population is reduced by any of a variety of factors that represent departures from a genetically ideal situation. These include the presence of nonbreeding individuals, skewed sex ratios, nonrandom tendencies of inbreeding or outcrossing, non-Poisson variation in progeny survivorship, the loss of genetic variability that may have occurred during previous periods of low population size (the bottleneck effect), and the tendency of the individuals of some species to mate with individuals from an area that may be smaller than that occupied by the population as a whole (the population structure effect). The cumulative effects of these factors are multiplicative and are usually expressed as the ratio of N_e/N. Between generations, N_e is the harmonic mean of the within-generation N_e's.

We can also calculate N_e for an entire metapopulation (Gilpin 1987). A very important variable is the rate of population turnover. In the absence of turnover, and with a minimum of one migrant per generation, the genetically effective size of the metapopulation is the sum of the N_e's of the local populations. At the other extreme of high turnover, N_e can be very low, since the entire set of living animals can be traced back to the few individuals who colonized a single patch. It is impossible to know the historical pattern of local turnover for the black-footed ferret and thus the extent to which its historical N_e may have been depressed by its metapopulation structure. One must, however, accept the possibility that the black-footed ferret is a species that has not been able to maintain high levels of genetic variation (Lacy, chap. 7, this volume).

Inbreeding depression results from the expression of deleterious genes as a result of the mating of close relatives. Most populations carry a "genetic load" of deleterious alleles. These alleles are usually rare and not particularly harmful in the heterozygous state; however, when closely related individuals mate, the chance of their offspring inheriting deleterious alleles in the homozygous state is increased. When deleterious alleles appear as homozygotes, they often result in the decline of such fitness traits as fertility, fecundity, and viability. If effective population size is large enough, natural selection can eliminate the individuals expressing the deleterious recessive alleles, and the effects on the population as a whole will be minimal. However, if the N_e is too small for selection to work effectively, the deleterious genes may become fixed in the population.

Furthermore, the deleterious effects of inbreeding can interact with random demographic events and result in even more severe threats to the persistence of small populations (Gilpin and Soulé 1986). For example, as population size becomes smaller, the incidence of inbreeding is likely to increase. As inbreeding increases, the incidence of reproductive failure is likely to increase as well. This results in a further reduction of population size, increasing inbreeding even more. Soon, a population can be caught in a downward spiral of ever-decreasing viability, generally leading to its extinction.

Groves and Clark (1986) calculated that a population size of 214 black-footed ferrets would be required to maintain an N_e of 50; thus, the N_e/N ratio is about 0.25. If the equilibrium number of breeding ferrets of Meeteetse is assumed to be 40, the genetically effective size of the population is about 10. At this size, the expected rate of increase in inbreeding in each generation (F) is 5%. This is 5 times the level at which natural selection for performance and fertility can balance inbreeding depression (Frankel and Soulé 1981).

Random changes in gene frequencies in populations are called genetic drift; these random changes sooner or later result in the loss of genetic variation, the rate of loss being dependent on population size. In small populations this loss can be quite severe and rapid. For example, in a population with an effective size of 10 individuals, only about 95% of the genetic variation in the population will be retained each generation. After 10 generations, if N_e's stay the same, 40% of the original genetic variability in the population will be lost. Losses of this magnitude can pose serious threats for short-term survival since the fitness of a population usually depends on the presence of sufficient variation in its gene pool to maintain normal fecundity and viability. Likewise, the adaptability of a population, or its ability to evolve, also depends on an even broader array of genetic resources in order to adjust to environmental change through the process of natural selection.

Genetic Variation and Metapopulation Structure in the Black-Footed Ferret

Kilpatrick et al. (1986) found no genetic variation in 3 salivary proteins sampled from 22 black-footed ferrets. Although this is a very small sample of loci, Simonsen (1982) found no variability at all at 16 loci in 6 other species of mustelids, suggesting that electrophoretic variation is minimal, if not absent, in this family (O'Brien et al., chap. 3, this volume). However, the absence of protein variation does not necessarily imply the absence of genetic variation for adaptively important characters (Hedrick et al. 1986). The accumulation by mutation of genetic variation affecting quantitative traits occurs much more quickly than it does for protein-coding loci (Lande 1976), because mutations that affect quantitative traits can occur at many loci. Nevertheless, because of its extreme rarity, it can be reasonably assumed that the black-footed ferret has lost some of the genetic variability that it had originally (Lacy, chap. 7, this volume).

Whatever levels of genetic variation were present in presettlement black-footed ferret populations were undoubtedly maintained in the species by low levels of migration among metapopulation units. In general, the number of genetically effective migrants that must be exchanged among adjacent subpopulations per generation to maintain effective panmixia is between 1 and 2 (Lande and Barrowclough 1987). Thus, rather low levels of gene flow can alleviate problems with inbreeding and loss of variability in subpopulations of a metapopulation system.

According to current theory in population biology, the risk of extinction of a population depends on the number of individuals the habitat can support (K), its per-capita birth rate, its rate of growth (r), the variability of its growth rate (V_i and V_e), and its genetically effective size (N_e). If a rare species can maintain a minimum breeding population size somewhere above the mid-tens to low hundreds (the exact number depending on its growth rate and per-capita birth rate), the odds favor its persistence, provided that there is no environmentally caused variation in these rates. If environmental variation affects birth and death processes in any significant way, however, a minimum viable population needs to be much larger to avoid extinction from demographic factors alone.

Even at population sizes that are adequate to buffer populations from extinction from the effects of V_i and V_e, low genetically effective sizes can cause additional problems. An N_e below 50 or so results in levels of inbreeding and loss of genetic variation that may also threaten short-term persistence. An N_e of around 500 is probably the minimum necessary for continuing evolution (Franklin 1980; Frankel and Soule 1981; Lande and Barrowclough 1987).

If the ferrets survive the current bottleneck, the population must be built up rapidly so that any further loss of genetic variation will be minimized. Any ferrets still extant in the wild should be captured to augment the gene pool in the captive-breeding colony. Furthermore, from what is known about the biology of the black-footed ferret, environmental variance appears to play a significant role in its population dynamics, and any recovery plan must recognize this important fact. When populations are reestablished in the wild, they must be located in reserves spaced far enough apart so that the effects of environmental variance will be minimized on the species as a whole. The management plan will have to include artificial movement of animals from reserve to reserve to facilitate recolonization following local extinctions and to alleviate the genetic problems associated with small, isolated populations.

In order to assure the potential for continuing evolution in the black-footed ferret, the target recovery level should aim for an N_e of around 500. The actual number of individuals necessary to maintain an effective population size this large is likely to be about 2,000. This implies a network of 20 reserves large enough to support 100 animals, or about 30 to 35 breeding adults each.

References

Forrest, S. C., et al. 1988. Population attributes for the black-footed ferret (*Mustela nigripes*) at Meeteetse, Wyoming, 1981–1985. *J. Mammal.* 69:261–73.

Frankel, O. H., and M. E. Soule. 1981. *Conservation and evolution.* Cambridge: Cambridge Univ. Press.

Franklin, I. R. 1980. Evolutionary change in small populations. In *Conservation biology: An evolutionary-ecological perspective*, ed. M. E. Soule and B. A. Wilcox, 135–49. Sunderland, Mass.: Sinauer Associates.

Gilpin, M. E. 1987. Spatial structure and population vulnerability. In *Viable populations for conservation*, ed. M. E. Soule, 125–40. Cambridge: Cambridge Univ. Press.

Gilpin, M. E., and M. E. Soule. 1986. Minimum viable populations: Processes of species extinction. In *Conservation biology: The science of scarcity and diversity*, ed. M. E. Soule, 19–34. Sunderland, Mass: Sinauer Associates.

Goodman, D. 1987a. The demography of chance extinction. In *Viable populations for conservation*, ed. M. E. Soule, 11–34. Cambridge: Cambridge Univ. Press.

Goodman, D. 1987b. Consideration of stochastic demography in the design and management of biological reserves. *Natural Resources Modeling* 1:205–34.

Groves, C. R., and T. W. Clark. 1986. Determining minimum population size for recovery of the black-footed ferret. *Great Basin Nat. Memoirs* 8:150–59.

Hedrick, P. W., et al. 1986. Protein variation, fitness, and captive propagation. *Zoo. Biol.* 5:91–99.

Kilpatrick, C. W., et al. 1986. Estimating genetic variation in the black-footed ferret—a first attempt. *Great Basin Nat. Memoirs* 8:145–49.

Lande, R. 1976. The maintenance of genetic variability in a polygenic character with linked loci. *Genet. Res. Camb.* 26:221–31.

Lande, R., and G. F. Barrowclough. 1987. Effective population size, genetic variation, and their use in population management. In *Viable populations for conservation*, ed. M. E. Soule, 87–124. Cambridge: Cambridge Univ. Press.

Leigh, E. G., Jr. 1981. The average lifetime of a population in a varying environment. *J. Theor. Biol.* 90:213–39.

MacArthur, R. H. 1972. *Geographical ecology.* New York: Harper and Row.

MacArthur, R. H., and E. O. Wilson. 1967. *The theory of island biogeography.* Princeton, N.J.: Princeton Univ. Press.

May, R. M. 1973. *Stability and complexity in model ecosystems.* Princeton, N.J.: Princeton Univ. Press.

Richter-Dyn, N., and N. S. Goel. 1972. On the extinction of colonizing species. *Theor. Pop. Biol.* 3:406–33.

Salwasser, H., et al. 1984. Wildlife population viability: A question of risk. *Trans. N. Am. Wildl. and Nat. Res. Conf.* 49:421–29.

Simonsen, V. 1982. Electrophoretic variation in large mammals. II. The red fox, *Vulpes vulpes*, the stoat, *Mustela erminea*, the weasel, *Mustela nivalis*, the beech marten, *Martes foina*, and the badger, *Meles meles*. *Hereditas* 96:299–305.

Soule, M. E., ed. 1987. *Viable populations for conservation.* Cambridge: Cambridge Univ. Press.

5

JONATHAN D. BALLOU

Inbreeding and Outbreeding Depression in the Captive Propagation of Black-Footed Ferrets

Captive propagation is a powerful tool in aiding the survival of endangered and threatened species (Conway 1980; Martin 1975). In recent years, new techniques for long-term genetic and demographic management have been developed (Dresser 1984; Flesness 1977; Foose 1983; Ralls and Ballou 1983). It is clear that in the case of the black-footed ferrets the priorities are demographic rather than genetic. As the captive population grows, however, genetic considerations will become more important.

This chapter discusses two genetic considerations important to the survival of the captive ferret population: inbreeding and outbreeding depression. Because both could significantly affect survival and reproduction in other captive populations, they should be addressed in the development of a black-footed ferret captive-breeding plan.

Inbreeding Depression

The deleterious effects of inbreeding on both reproductive and survival characteristics have been well documented in a large variety of domestic, laboratory, and captive populations. Wright's (1922) analysis of the effects of inbreeding in guinea pigs probably remains the most extensive study of inbreeding effects to date. During 18 years of experimental inbreeding of guinea pig colonies for the Department of Agriculture, Wright (1977) documented reduction in size and frequency of litters, a decrease in the proportion of animals born alive and reared to 3 months of age, and a decrease in birth weight and weight at 3 months of age (Wright 1977). Similar results have been shown in cattle (Young et al. 1969), pigs (McPhee et al. 1931), rabbits (Chai 1969), dogs (Wildt et al. 1982), and many other domestic animals (Ralls and Ballou 1983).

Few data are available on the effects of inbreeding in mustelids. Although inbreeding in domestic ferrets is a concern (Marshall, pers. comm.), data on the effects are not available. Johansson (1961) conducted inbreeding experiments on ranch-bred Swedish mink (*Mustela vison*). Mink ranches in Sweden have traditionally developed their mink populations from small numbers of wild-caught animals. This approach, in addition to selective breeding for coat color, has resulted in high levels of inbreeding in the ranch mink populations. Crossings between different strains, however, may have restored levels of heterozygosity. Despite the previous inbreeding in the population, additional inbreeding of these ranch-bred mink still resulted in reduced survivorship, reproduction, and litter size and was severe enough to cause extinction by the fourth generation in the experimental inbred lines (Fig. 5.1).

Information on the effects of inbreeding in nondomestic species

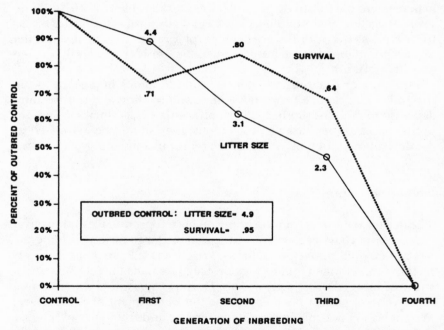

Figure 5.1. The decrease in survival and litter size in inbred mink. The mean effect of inbreeding in 8 lines of inbred mink at 4 mink ranches over 4 generations is shown as a percentage of the outbred control. Lines at 3 of the 4 ranches went extinct after the 2d generation, and the last lines survived only to the 3d. The number of offspring produced by the outbred controls and the 1st, 2d, and 3d inbred generations were 2,549, 157, 99, and 25, respectively (Johansson 1961).

comes primarily from breeding and pedigree records of captive populations. Ralls and her colleagues (1983) summarized the effects of inbreeding on juvenile and infant mortality in 44 mammalian populations
from 7 orders, 21 families, and 36 genera. Forty-one of the 44 populations had higher inbred than noninbred juvenile mortality ($p = 0.001$;
Fig. 5.2). Significant differences between inbred and noninbred mortality rates within populations were found in only 14 (31%) of the populations. Small sample sizes of most of the populations surveyed limited
the statistical power of analyses (see below). Nevertheless, the results are
clearly indicative of a general trend in the deleterious effects of inbreeding on captive mammalian populations. Numerous other studies support these conclusions (Templeton and Read 1983; Shoemaker 1982;
Buisman and van Weeren 1982; Roberts 1982).

Studies on the effects of inbreeding in natural populations are scarce.
Packer (1979) found reduced survival in baboon offspring from related
parents; and evidence for inbreeding depression has been found in the
great tit (Bulmer 1973).

Data from studies correlating the level of inbreeding on mortality and

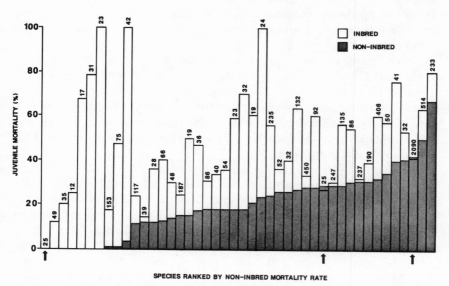

SPECIES RANKED BY NON-INBRED MORTALITY RATE

Figure 5.2. Inbred (shaded) and noninbred mortality rates in 44 different mammalian
populations (Ralls and Ballou 1983). The populations are ranked by their noninbred
mortality rates. The numbers represent total sample sizes in each population. The arrows
represent the 3 populations that show higher (nonsignificant) noninbred mortality rates
(from left to right: four-striped rat, kudu, and pig-tailed macaque).

reproductive characteristics in domestic species show an approximate 10 to 20% decrease in fitness with each 10% increase in inbreeding. Ralls et al. (1988) examined the correlation between inbreeding coefficients and mortality in 40 different captive populations. The mortality rates associated with different levels of inbreeding were calculated and regressed on the inbreeding coefficients within each of the populations. This method was analogous to that used by Templeton and Read (1984) in analyzing inbreeding effects in Speke's gazelle (*Gazella spekei*). The severity of the inbreeding effect (the number of lethal equivalents [Morton et al. 1956]) could then be estimated by the rate at which mortality increased with the increasing inbreeding coefficient (the slope of the regression). In addition, a predicted mortality rate for any specified level of inbreeding could be calculated using the estimated regression equation.

Figure 5.3 illustrates the distribution of the predicted increase in mortality associated with an inbreeding coefficient of 10% for the 40 populations surveyed (Ralls et al. 1988). Although the effect of inbreeding on early mortality was highly variable, a 10% increase in inbreeding caused an average 10% decrease in survival; 50% of the populations had predicted decreased levels of survival between 3% and 15%. Figure 5.4 shows the predicted relationship between inbreeding coefficients and survivorship for the median and the upper and lower quartiles of the distribution.

These figures probably underestimate the inbreeding effects that occur in wild populations. In captivity, weak inbred animals that benefit from veterinary care might be expected to have lower mortality rates than inbred animals in the wild. Nevertheless, these results illustrate the potential effect inbreeding can have on mortality rates in captive mammal populations, as well as the variation in responses to inbreeding within different populations.

The loss of heterozygosity due to inbreeding can be directly correlated with fitness components such as survival and reproduction in captive populations with known pedigrees. In populations without pedigrees, evaluating the potential effects of the loss of heterozygosity is much more difficult. Evidence of inbreeding depression in these populations is indirectly inferred from studies on the relationship between heterozygosity and fitness. If such a positive relationship does exist, the loss of heterozygosity due to inbreeding, genetic drift, or other factors will result in a population with lower fitness.

Several studies show that there is a general, though not universal, positive relationship between heterozygosity and various fitness components (survivorship, disease resistance, growth and developmental rate,

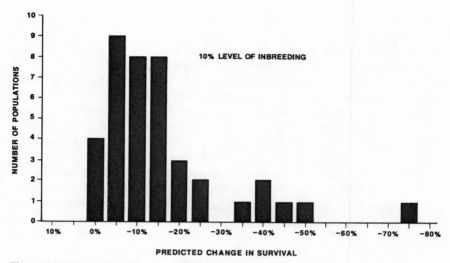

Figure 5.3. The predicted change in survival associated with an inbreeding level of 0.10 in 40 mammalian populations (Ralls et al. 1988).

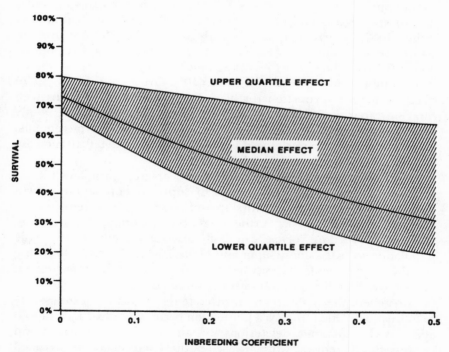

Figure 5.4. The predicted relationship between inbreeding and mortality in 40 mammalian populations (Ralls et al. 1988). The median, upper, and lower quartile effects of the distribution of 40 populations are shown.

reproductive rates, and developmental stability) in invertebrate and vertebrate populations (Allendorf and Leary 1986; Mitton and Grant 1984). Although Allendorf and Leary (1986,73) discuss several minor interpretational problems and complications common to many of these studies, they conclude that "there are many well-documented examples of advantages of heterozygotes in several components of fitness. There are fewer examples of disadvantages . . . Thus we conclude that, in general, there is a positive relationship between heterozygosity and fitness."

Inbreeding data from domestic animals and captive populations, as well as the evidence for a general relationship between heterozygosity and fitness, indicate that inbreeding and the loss of genetic diversity can have potentially disastrous effects on the survival of small populations near extinction. This can be illustrated for ferrets by observing the result of additional mortality due to inbreeding on estimates of black-footed ferret minimum viable population (MVP) sizes (Brussard and Gilpin, chap. 4, Harris et al., chap. 6, Gilpin and Soule 1986). Although the development of MVP estimates is beyond the scope of this chapter, a consideration in such estimates is the amount of time it takes a population of specified characteristics to go to extinction (MacArthur and Wilson 1967). A ferret population of effective size 5 will have an average 10% increase in inbreeding each generation. Data from captive populations (Fig. 5.4) show that the average increase in mortality associated with an inbreeding level of 10% is also 10%. Groves and Clark (1986) calculate times-to-extinction for black-footed ferret populations with carrying capacities of 40 and 50 (Table 5.1). This table conveniently illustrates the effects of increasing the death rates (u) in 10% increments. Times-to-extinction are reduced exponentially with each 0.10 increase in u, or each 10% increase in inbreeding.

Increased inbreeding and the decrease of heterozygosity, with its associated inbreeding depression, act to suppress population growth through a reduction in survivorship and fecundity. The reduction in growth and population size further increases the amount of inbreeding. Such an "inbreeding vortex" can easily drive an already small, fragile population to extinction (Gilpin and Soule 1986). Clearly, inbreeding and the loss of genetic diversity must be considered in the management of a captive breeding program for black-footed ferrets.

Two genetic mechanisms are involved in the reduction of fitness due to inbreeding. One is that as a population becomes increasingly inbred and more homozygous, deleterious recessive alleles are unmasked and expressed. The reduced fitness caused by the deleterious alleles is called "mutational load." The second mechanism is overdominance, or hetero-

Table 5.1.   Effects of Increased Death Rates
on the Time-to-Extinction in Black-Footed
Ferret Populations

Carrying Capacity	Birth-rate	Death Rate	Time-to-Extinction
	.5	.2	1.0×10^{15}
40	.5	.3	3.0×10^{7}
	.5	.4	1.0×10^{3}
	.5	.2	8.0×10^{18}
50	.5	.3	5.0×10^{9}
	.5	.4	1.0×10^{4}

Source: Groves and Clark 1986.
Note: The implications of an additional 10% mortality due to inbreeding on the survival of the ferret populations can be seen as an exponential decline in the time-to-extinction as death rates increase.

zygote superiority. Heterozygous loci may, for a variety of reasons, have higher fitness than homozygous loci (Mitton and Grant 1984). Such loci are called "heterotic" loci. In inbred populations, there is a reduction in the frequency of heterozygous loci and a reduction in any associated benefit. This type of reduction in fitness is called "segregational load." Both mutational and segregational load contribute to inbreeding depression, with the depression associated with deleterious alleles being perhaps the more important of the two (Allendorf and Leary 1986).

The severity of inbreeding depression is a function of the amount of mutational and segregational load carried by that population. Populations clearly vary in the amount of load they carry (Fig. 5.2), as evidenced by the large variety of responses to inbreeding in the populations surveyed by Ralls et al. (1988).

Populations with a previous history of inbreeding might be expected to carry less of a mutational load than outbred populations. In the absence of genetic drift, selection against deleterious alleles will quickly reduce the mutational load with even slight levels of inbreeding. However, segregational load will be reduced at a slower rate and remain fixed in populations that become homozygous. Outbreeding populations accumulate both mutational and segregational load and have high levels of inbreeding depression when inbred. On the other hand, naturally inbreeding populations or populations that have passed through bottlenecks might be expected to have low levels of inbreeding depression when further inbred. The lack of inbreeding depression in populations that have a history of inbreeding is documented in domestic (Lasley 1978) and laboratory (Musialek 1980) animal populations, as well as in studies of human populations (Rao and Inbaraj 1980).

A viable approach to population management, therefore, might be to purge the genetic load from populations not previously exposed to inbreeding. Such a program, however, could have disastrous consequences on the survival of small, endangered populations.

The cost of successfully purging a population of its genetic load can be prohibitive. The increased mortality and reduced fecundity incurred during the process of exposing the mutational and segregational loads to selection in small, inbred lines are often severe enough to cause extinction. In efforts to produce successful inbred lines, Bowman and Falconer (1960) found that only 1 of 20 lines of mice survived intense inbreeding. The same results have been shown repeatedly in both laboratory populations (Lorenc 1980) and populations founded with wild-caught individuals (Lynch 1977; Soule 1980). The odds of successfully purging the genetic load from a single, small population of an endangered species are dangerously low.

Additionally, inbred lines that do survive are often less fit than their outbred counterparts. This is because the segregation load may remain in the inbred, homozygous population, even though the mutational load is absent. Reproduction and survival may be reduced, and the population may become much more susceptible to outbreaks of disease and environmental stochasticity. A case in point is the cheetah (*Acinonyx jubatus*). O'Brien et al. (1983, 1985) found that the cheetah has extremely low levels of genetic variation and suggested that the cheetah has had a history of small population size and is highly inbred. In addition, cheetahs are notoriously difficult to breed in captivity, have high levels of abnormal sperm, and appear to be highly sensitive to disease (O'Brien et al. 1985). The problem may be even more severe in very small populations subject to genetic drift. In small populations, genetic drift has more effect on changes in allele frequencies than does selection; and deleterious alleles, with their associated reduction in fitness, can become fixed in these small populations.

Intentional selection against deleterious alleles also drastically changes the genetic characteristics of the population. This domestication is clearly undesirable in populations intended for future release into natural habitats. The intentional purging of genetic load is indeed a risky option with certain undesirable results and should be carefully considered before being used in the management of endangered species (Hedrick et al. 1986).

It would seem, however, that one could safely ignore the threat of inbreeding depression in naturally inbreeding populations or populations with a history of small population size since these populations might be expected to be purged already of their genetic load. The degree to which a population is purged is a function of both the selection acting on a population and the degree and rate of inbreeding in the population. Both selection and the degree of inbreeding (in wild populations) are extremely difficult to estimate, as are their effects, and are often modeled using a variety of assumptions. The results are indicative of a range of possible scenarios (Lacy and Clark, chap. 7, this volume; Ralls et al. 1983). Additionally, it is not always true that historically inbred populations or populations with low levels of detectable isozyme variation show no depression with additional inbreeding. Absence of electrophoretically detectable protein variations does not necessarily imply total absence of genetic variation for adaptively important characteristics (Hedrick et al. 1986). Despite a history of domestication and inbreeding, populations may still be susceptible to inbreeding depression (recall the Swedish mink [Johansson 1961]).

The Père David's deer (*Elaphurus davidianus*) was first brought to the attention of the West by the missionary Father David, who in 1865 noticed this unusual deer in the hunting compound of the Chinese emperor. Extinct in the wild, Père David's deer had probably bred in captivity for as many as 3,000 years before being rediscovered. Father David eventually managed to send a few deer to Europe (exactly how many is unknown, but probably less than 5 lived to reproduce). The current population has descended from the captive herd at Woburn Abbey, which itself was founded by fewer than 18 deer from the European population. Additional bottlenecks occurred during World War II. The history of Père David's deer is characterized by a series of bottlenecks, and the species is believed to be highly inbred (Whitehead 1978; Foose and Foose 1983). Initial electrophoretic analysis confirms this (Ryder et al. 1981). Nevertheless, Foose (1983) found significantly higher mortality rates in inbred Père David's deer young than in noninbred young.

Additionally, even though the cheetah is almost completely lacking in genetic variation (O'Brien et al. 1983, 1985), initial results suggest that captive inbred cheetahs suffer from significantly higher mortality rates than noninbred cheetahs (O'Brien et al. 1985). A more detailed analysis of inbreeding and mortality patterns in both of these species will be necessary to confirm this association.

These data suggest that one should take care in making assumptions regarding the severity of inbreeding depression in populations with a history of small population size or low levels of genetic variation. It cannot automatically be assumed that inbreeding problems can be ignored in these populations.

The extant population of black-footed ferrets at Meeteetse is limited to approximately 3,000 hectares, and historical records indicate that the population has been both small (less than 100 animals) and probably genetically isolated since the 1930s (Clark et al. 1986; Lacy and Clark, chap. 7, this volume). Given a generation time of approximately 1.5 years, Lacy and Clark estimate that the Meeteetse population has been in a bottleneck for over 30 generations and could have lost between 16 and 73% of the variation present in the 1930s, which may itself have been very low. The historical bottleneck and the ensuing loss of variation have led several to suggest that the population has already been purged of its genetic load and that the genetic consequences of additional inbreeding can be ignored in developing captive-breeding plans (Pettus 1985).

Given the uncertainty of our assumptions regarding the historical events in the population and their effects, as well as the consequences of ignoring the potential impact of inbreeding depression, it seems that such advice is unsound. Consideration of the potential effects of inbreeding depression should be part of an integrated captive-breeding plan for black-footed ferrets.

Outbreeding Depression

A second consideration in developing a captive-breeding program for black-footed ferrets is the potential for outbreeding depression. Outbreeding depression can be defined as a reduction in fitness due to crossing individuals that are too distantly related. One type of outbreeding depression results from crossbreeding individuals from populations that have adapted to different local environments (local adaptation). Crossing individuals between these regions may then result in offspring fit for neither of the parents' habitats.

As an example of outbreeding depression due to local adaptation, Templeton et al. (1986) refer to translocation efforts with ibex. To supplement a previously introduced population of Tatra mountain ibex (*Capra ibex ibex*) in Czechoslovakia, bezoars (*C. i. aegrarus*) and Nubian ibex (*C. i. nubiana*) were imported from Turkey and the Sinai. Offspring of the subspecific hybrids were born in the dead of winter and could not survive. As a result, the population went extinct over several years (Greig 1979).

Outbreeding depression can also result from disruption of coadapted gene complexes. Coadapted gene complexes can occur when a genetic or karyotype complex evolves or adapts in response to other genetic complexes or chromosomes. This differs from local adaptation in that genetic complexes are adapting to the state of other genes rather than to local environments. An example of karyotypic coadaptation can be found in the owl monkey (*Aotus trivirgatus*). Owl monkeys have several different chromosomal races (de Boer 1982), and reproductive fitness is improved when individuals with similar chromosome forms are paired in captivity (Cicmanec and Campbell 1977; Elliott et al. 1976).

Templeton et al. (1986) discuss techniques for predicting the possibility of outbreeding problems. Evidence of large gene-frequency differences between populations may be indicative of potential outbreeding problems. However, such differences may not necessarily be the result of coadaptation. In very small isolated populations, evolution is strongly determined by genetic drift, and differentiation may be due to drift rather than local adaptation or coadaptation due to selection. Evidence of gene-frequency differences between large populations not subject to genetic drift is of more weight in detecting the potential for outbreeding depression. Obviously the biogeography and history of any future black-footed ferret population will have to be examined to interpret any gene-frequency differences found.

Individuals can be genetically screened for karyotypic differences. Such differences are probably the best indicator of potential outbreeding problems. Differences in gene frequencies or karyotypes are only suggestive of potential problems and should not be interpreted as absolute indicators of an outbreeding effect. Outbreeding depression has been found in populations without detectable differences, while populations with considerable genetic differences have been found to have no outbreeding depression (Templeton et al. 1986).

Outbreeding depression has been found in plants, invertebrates, and, to a more limited extent, in vertebrate species (Shields 1982; Templeton 1986). Many studies on vertebrates indicate variable reactions to outcrossing (Cade 1983). However, evidence for outbreeding effects in mammals is scarce. Initial studies in captive populations indicate that outbreeding depression is not a contributing factor to mortality in these populations (Templeton and Read 1984). Inbreeding depression seems to be more of a problem than outbreeding depression and, unless genetic assays indicate otherwise, should be considered a higher priority than outbreeding depression.

The question of how important local adaptations and coadapted gene complexes are to the future management of the black-footed ferret is of interest only if other isolated populations of ferrets are found. Dispersal patterns of the ferrets suggest that different local adaptations or genetic coadaptation have probably not evolved within the Meeteetse population.

Management options for populations showing signs of inbreeding and outbreeding depression have been discussed in general terms elsewhere (Foose et al. 1986; Templeton et al. 1986; Templeton and Read 1984). Basically, outbreeding is a suggested management plan for populations likely to suffer from inbreeding depression. Similarly, populations likely to suffer from outbreeding depression should be managed as contained groups (Templeton et al. 1986).

These two recommendations are, in a sense, diametrically opposed in very small populations: avoiding inbreeding and maintaining genetic diversity necessitates outbreeding, while avoiding outbreeding necessitates population segregation and thus increased inbreeding. When managing small populations, we must therefore initially decide which strategy to adopt, since we cannot implement a management plan to avoid both outbreeding and inbreeding.

If the captive population of black-footed ferrets is founded solely from the Meeteetse population, as is now the case, it is unlikely that outbreeding depression will be a problem in the captive population. Although it is also possible that the population has been purged of its genetic load due to its history of moderate previous inbreeding (Lacy and Clark, chap. 7, this volume), conservative management practices dictate recognition of potential inbreeding effects.

Frankham et al. (1986) and Foose et al. (1986) discuss a variety of goals for the management of captive populations. Although the maintenance of genetic diversity is considered the most desirable goal for captive-breeding plans, they recognize the need to relax genetic considerations in favor of demographic considerations when the survival of the population is threatened. Certainly, this was the situation with the initial 18 black-footed ferrets in captivity. Strict emphasis should be placed on increasing the number of animals as rapidly as possible. This is necessary for both demographic reasons (small populations are more susceptible to extinction [Gilpin and Soule 1986]) and genetic reasons (the bottleneck effects of a founding event are minimized if postbottleneck population growth is rapid [Nei et al. 1975; Denniston 1978]). During this critical phase of growth, genetic considerations associated with maintaining genetic diversity (such as avoiding inbreeding and managing founder contribution [Foose et al. 1986]) must be secondary and should be considered only when the growth of the population is not compromised. When searching for reproductively compatible pairs, however, pairings between unrelated animals should be considered before those between related individuals.

During the critical phase, the effects of inbreeding should be carefully monitored. Records on successful and unsuccessful pairings, mortality data, parentage information, and other studbook data should be kept for future analysis. As the population becomes larger and more animals are available for breeding, there will be more flexibility in implementing genetic considerations. At this point (the "intensive care" phase), it is appropriate to begin evaluating the effects of inbreeding using whatever data have already been compiled during the critical phase of the population.

Various methods of evaluating the effects of inbreeding on mortality and reproduction are described elsewhere (Ralls et al. 1979; Templeton and Read 1984; Lee 1980). It is possible, however, that not enough data will have been produced to make a proper evaluation. A central problem in detecting inbreeding effects is the amount of data necessary to make a statistically valid comparison between inbred and noninbred mortality rates. The concern is that a negative result might be misinterpreted. One might conclude that there is no inbreeding effect if a nonsignificant difference is found, when, in fact, not enough data have been collected to detect an effect. The power of this test must be high enough to draw a proper conclusion.

Table 5.2 shows the sample sizes required to detect various differences between inbred and noninbred mortality rates with 90 and 95% confidence (Lachin 1981). For example, to be 90% certain of detecting an inbreeding depression of 40% relative to a baseline (noninbred) mortality of 30% (inbred mortality of 70%) requires a total sample size of 50 (25 inbred and 25 noninbred births). Even larger sample sizes will be required if most births are inbred, which is likely to be the case. Of course, the larger the inbreeding effect, the fewer data are needed to detect it; and it is possible that inbreeding effects of the magnitude likely to cause severe problems for a population will be easily detected early in its history.

If there are not enough data to detect the level of inbreeding effects that might be of concern to the population, or if data are available and such an effect is slight, then it is recommended that the population continue to be managed to enhance population growth and maintain genetic diversity. Data on the effects of inbreeding should continue to be monitored in either case. If an inbreeding effect becomes strong enough to significantly affect mortality patterns, it should be recognized soon.

If an inbreeding effect is detected and appears to be strong enough to affect the overall health and future chances of survival of a population,

Table 5.2. Sample Sizes Required to Detect Different Degrees of Inbreeding Depression

Inbred Mortality Rate	Noninbred (Base) Mortality Rate								
	0.1	0.2	0.3	0.4	0.5	0.6	0.7	0.8	0.9
A: Sample required to detect a difference with 90% confidence									
0.1	—								
0.2	398	—							
0.3	126	639	—						
0.4	65	176	776	—					
0.5	40	83	202	844	—				
0.6	27	48	90	210	844	—			
0.7	19	30	50	90	202	776	—		
0.8	13	20	30	48	83	176	639	—	
0.9	9	13	19	27	41	68	133	433	—
1.0	6	9	12	16	22	30	45	73	159
B: Sample required to detect a difference with 95% confidence									
0.1	—								
0.2	507	—							
0.3	159	806	—						
0.4	82	222	980	—					
0.5	50	104	254	1,066	—				
0.6	33	59	114	265	1,066	—			
0.7	23	37	62	114	254	980	—		
0.8	16	24	37	59	104	222	806	—	
0.9	11	16	23	34	51	85	168	547	—
1.0	7	10	14	19	27	38	56	92	200

Source: Lachin 1981.
Note: The sample sizes specified are the total number of births equally distributed between inbred and noninbred young required to detect a significant difference at the 0.05 level with a power of 0.90 (A) and 0.95 (B).

then a management plan to purge the population of its genetic load might be considered.

Templeton and Read (1984) discuss a management plan applied to a captive population of Speke's gazelle suffering from strong inbreeding depression. Their goal was to eliminate inbreeding depression by adapting the population to inbreeding. This was accomplished by selectively breeding healthy, inbred animals who would produce inbred offspring. Care was taken to avoid producing extremely inbred offspring as well as to choose parents that would maximize the genetic viability of the gene pool. Using the above management recommendations, they reduced the genetic load in the population by almost one-half in a 3-year period (Templeton and Read 1984). Such a management plan could be considered for the black-footed ferret population if significant inbreeding depression was found.

Implementing a plan to purge the population of its genetic load, however, must be seriously considered because of the severe disadvantages. These are particularly serious since release of captive-born animals to establish and supplement existing populations is a primary goal of the black-footed ferret recovery plan (Richardson et al. 1986). Such disadvantages seem negligible, however, if the alternative is the extinction of the captive population and possibly the entire species.

Given the consequences of implementing a plan to purge genetic load, how severe does inbreeding depression have to be before such a plan is recommended? I recommend that it be used only if the inbreeding is severe enough to jeopardize the continued survival of the population. This condition would exist if the depression was severe enough to: (a) keep the population small enough to remain dangerously susceptible to demographic fluctuations and catastrophic events, or (b) cause a significant declining trend towards extinction. Either of these conditions might be considered sufficient to implement a plan of reducing or purging the population of its genetic load.

Once the levels of genetic load are reduced, the management plan could then revert back to maintaining genetic diversity. Templeton and Read's (1984) results suggest that populations can be purged of their genetic load relatively quickly. In fact, a plan alternating between purifying selection and maintaining genetic diversity could also be considered.

Managing Inbreeding and Outbreeding Depression

The possibility of discovering other black-footed ferret populations raises some interesting questions regarding the incorporation of new animals into a captive-breeding program. There would be considerable genetic and demographic advantages in supplementing the Meeteetse-founded captive population with animals from other populations. A larger founder size will enhance the overall maintenance of genetic diversity, as well as increase the probability of successful reproduction and rapid growth. Merging the population could increase levels of heterozygosity, thereby restoring heterotic loci and promoting increased fitness in the offspring. If one chose to ignore the potential effect of outbreeding depression, the population should be initially merged and perhaps later subdivided, while maintaining limited gene flow (Foose et al. 1986).

Other populations that might be found will likely have been isolated from the Meeteetse population for many generations and may be fixed for alleles different from those in the Meeteetse population. It is possible that any fixed genetic difference will be due to genetic drift rather than selection. Under these conditions, the potential for outbreeding does exist and should be considered in the development of a breeding plan. However, inbreeding considerations should still take priority over outbreeding considerations.

If only a few individuals from another population are found, the primary goal will again be population growth, with the emphasis on choosing reproductively compatible pairings rather than genetically compatible pairings. However, a choice must first be made whether to search for reproductively compatible mates within or between populations. The karyotypes of the 2 populations should be compared. If differences are found, individuals within populations should be preferentially mated. If no differences exist, individuals between populations should be preferentially mated. As the population grows and there is more freedom in mate selection, the data can be analyzed for inbreeding and outbreeding effects using the methods described by Templeton and Read (1984). If inbreeding depression is a severe problem, preference should be given to outcrossing individuals rather than implementing a plan to purge the population of its genetic load. Likewise, if outbreeding depression is severe, plans should be developed to avoid pairings of individuals between populations (but also to avoid pairings between related individuals). If the data are inconclusive, then the population should be allowed to continue to grow. Pairings should be selected on the basis of maximizing genetic diversity. Test inbred and outbred crosses could be made if the population grows large enough so that animals become available. Additionally, if other large populations of ferrets are found, test pairings between populations could be conducted sooner.

If such pairings are compatible, a decision to merge the captive populations totally, partially, or not at all would have to be made. If inbreeding depression is absent, the best strategy might be to establish a third captive population by outcrossing some individuals from each of the 2 original populations, as suggested by Lacy and Clark (chap. 7, this volume).

This discussion of the potential outbreeding and inbreeding effects within and between future captive populations of black-footed ferrets should serve primarily to illustrate the types of management options that might be considered. Obviously, it is impossible to predict the characteristics of either the future captive population or any other yet-to-be-discovered black-footed ferret populations. Management decisions will have to be based on the specific genetic and demographic characteristics of the populations present at that time.

References

Allendorf, F. W., and R. F. Leary. 1986. Heterozygosity and fitness in natural populations of animals. In *Conservation biology: The science of scarcity and diversity*, ed. M. E. Soule, 57–76. Sunderland, Mass.: Sinauer Assoc.

Bowman, J. C., and D. S. Falconer. 1960. Inbreeding depression and heterosis of litter size in mice. *Genet. Res. Camb.* 1:262–74.

Buisman, A. K., and R. van Weeren. 1982. Breeding and management of Przewalski horses in captivity. In *Breeding Przewalski horses in captivity for release into the wild*, ed. J. Bouman, I. Bouman, and A. Groeneveld, 76–111. Rotterdam: Foundation for the Preservation and Protection of Przewalski Horses.

Bulmer, M. G. 1973. Inbreeding in the great tit. *Hereditas* 30:313–25.

Cade, T. J. 1983. Hybridization and gene exchange among birds in relation to conservation. In *Genetics and conservation*. ed. C. M. Schonewald-Cox, S. M. Chambers, B. MacBryde, and L. Thomas, 288–309. Menlo Park, Calif.: Benjamin-Cummings.

Chai, C. K. 1969. Effects of inbreeding in rabbits. *J. Hered.* 60:64–70.

Cicmanec, J. C., and A. K. Campbell. 1977. Breeding the owl monkey (*Aotus trivirgatus*) in a laboratory environment. *Lab. Anim. Sci.* 27:512–17.

Clark, T. W., et al. 1986. Description and history of the Meeteetse black-footed ferret environment. *Great Basin Nat. Memoirs* 8:72–84.

Conway, W. G. 1980. An overview of captive propagation. In *Conservation biology*, ed. M. E. Soule and B. A. Wilcox, 199–208. Sunderland, Mass.: Sinauer Assoc.

de Boer, L. E. M. 1982. Karyological problems in breeding owl monkeys (*Aotus trivirgatus*). *Internat. Zoo Yearbook* 22:119–24.

Denniston, C. 1978. Small population size and genetic diversity: Implications for endangered species. In *Endangered birds: Management techniques for preserving threatened species*, ed. S. A. Temple, 281–89. Madison, Wis.: Univ. of Wisconsin Press.

Dresser, B. 1984. Proceedings of the conference on reproductive strategies for endangered wildlife. *Zoo Biol.* 3:307–9.

Elliott, M. W., et al. 1976. Management and breeding of *Aotus trivirgatus*. *Lab. Anim. Sci.* 26:1037–40.

Flesness, N. 1977. Gene pool conservation and computer analysis. *Internat. Zoo Yearbook* 17:77–81.

Foose, T. J. 1983. The relevance of captive propagation to the conservation of biotic diversity. In *Genetics and conservation*, ed. C. M. Schonewald-Cox,

S. M. Chambers, B. MacBryde, and L. Thomas, 374–401. Menlo Park, Calif.: Benjamin-Cummings.

Foose, T. J., and E. Foose. 1983. Demographic and genetic status and management. In *The biology and management of an extinct species: Père David's deer*, ed. B. B. Beck and C. Wemmer, 133–86. Park Ridge, N.J.: Noyes.

Foose, T. J., et al. 1986. Propagation plans. In *Proc. workshop genetic management of captive populations*, ed. K. Ralls and J. D. Ballou. *Zoo Biol.* 5:139–46.

Forrest, S. C., et al. 1988. Population attributes for the black-footed ferret (*Mustela nigripes*) at Meeteetse, Wyoming, 1981–1985. *J. Mammal.* 69:261–73.

Frankham, R., et al. 1986. Selection in captive populations. In *Proc. workshop genetic management of captive populations*, ed. K. Ralls and J. D. Ballou. *Zoo Biol.* 5:127–38.

Gilpin, M. E., and M. E. Soule. 1986. Minimum viable populations: Processes of species extinction. In *Conservation biology: The science of scarcity and diversity*, ed. M. E. Soule, 19–34. Sunderland, Mass.: Sinauer Assoc.

Greig, J. C. 1979. Principles of genetic conservation in relation to wildlife management in Southern Africa. *South African J. Wildl. Res.* 9:57–78.

Groves, C. R., and T. W. Clark. 1986. Determining minimum population size for recovery of the black-footed ferret. *Great Basin Nat. Memoirs* 89:150–59.

Hedrick, P. W., et al. 1986. Protein variation, fitness and captive propagation. In *Proc. workshop genetic management of captive populations*, ed. K. Ralls and J. D. Ballou. *Zoo Biol.* 5:91–99.

Johansson, I. 1961. Studies on the genetics of ranch bred mink. I. The results of an inbreeding experiment. *Z. Tierzuecht. Zuechtungs-Biol.* 72:293–97.

Lachin, J. M. 1981. Introduction to sample size determination and power analysis for clinical trials. *Controlled Clinical Trials* 2:93–113.

Lacy, R. C. 1987. Loss of genetic diversity from managed populations: interacting effects of drift, mutation, immigration, selection and population subdivision. *Conserv. Biol.* 1:143–58.

Lasley, J. F. 1978. *Genetics of livestock improvement*. Englewood Cliffs, N.J.: Prentice-Hall.

Lee, E. T. 1980. *Statistical methods for survival data analysis*. Belmont, Cal.: Lifetime Learning Pubns.

Lorenc, E. 1980. Analysis of fertility in inbred lines of mice derived from populations differing in their genetic load. *Zeierzeta Laboratoryjne* 17:3–16.

Lynch, C. B. 1977. Inbreeding effects upon animals derived from wild populations of *Mus musculus*. *Evolution* 31:525–37.

MacArthur, R. H., and E. O. Wilson. 1967. *The theory of island biogeography*. Princeton, N.J.: Princeton Univ. Press.

McPhee, H. C., et al. 1931. An inbreeding experiment with Poland China swine. *J. Hered.* 22:383–403.

Martin, R. D. 1975. *Breeding endangered species in captivity*. London: Academic Press.

Mitton, J. B., and M. C. Grant. 1984. Associations among protein heterozygosity, growth rate, and developmental homeostasis. *Ann. Rev. Ecology and Syst.* 15:479–500.

Morton, N. E., et al. 1956. An estimate of the mutational damage in man from data on consanguineous marriages. *Proc. Nat. Acad. of Sci.* 42:855–63.

Musialek, B. 1980. Effect of the population size on decrease of fertility in mice. *Genetica Polonica* 21:461–75.

Nei, M., et al. 1975. The bottleneck effect and genetic variability in populations. *Evolution* 29:1–10.

O'Brien, S. J., et al. 1983. The cheetah is depauperate in genetic variation. *Science* 221:459–62.

O'Brien, S. J., et al. 1985. Genetic basis for species vulnerability in the cheetah. *Science* 227:1428–34.

Packer, C. 1979. Inter-troop transfer and inbreeding avoidance in *Papio anubis*. *Anim. Behav.* 27:1–36.

Pettus, D. 1985. Genetics of small populations. In *Black-footed ferret workshop proceedings, Wyoming Game Dept.*, ed. S. Anderson and D. Inkley, 22.1–22.11. Cheyenne, Wyo.: Wyoming Fish and Game Dept.

Ralls, K. and J. D. Ballou. 1983. Extinction: Lessons from zoos. In *Genetics and conservation.* ed. C. M. Schonewald-Cox, S. M. Chambers, B. MacBryde, and L. Thomas, 164–84. Menlo Park, Calif.: Benjamin-Cummings.

Ralls, K., et al. 1983. Genetic diversity in California sea otters: Theoretical considerations and management implications. *Biol. Conserv.* 25:209–32.

Ralls, K., et al. 1979. Inbreeding and juvenile mortality in small populations of ungulates. *Science* 206:1101–3.

Ralls, K., J. Ballou, and A. R. Templeton. 1988. Estimates of lethal equivalents and the cost of inbreeding in mammals. *Conservation Biology* 2:185–93.

Rao, P. S. S., and S. G. Inbaraj. 1980. Inbreeding effects on fetal growth and development. *J. Med. Genet.* 17:27–33.

Richardson, L., et al. 1986. Black-footed ferret recovery: A discussion of some options and considerations. *Great Basin Nat. Memoirs* 8:169–84.

Roberts, M. 1982. Demographic trends in a captive population of red pandas (*Ailurus fulgens*). *Zoo Biol.* 1:119–26.

Ryder, O. A., et al. 1981. Monitoring genetic variation in endangered species. In *Evolution Today*, ed. G. G. E. Scudder and J. L. Reveal, 417–24. Pittsburgh: Carnegie-Mellon Univ.

Shields, W. M. 1982. *Philopatry, inbreeding and the evolution of sex.* Albany, N.Y.: State Univ. of New York.

Shoemaker, A. H. 1982. The effect of inbreeding and management on propagation of pedigree leopards. *Internat. Zoo Yearbook* 22:198–206.

Soule, M. E. 1980. Thresholds for survival: Maintaining fitness and evolutionary potential. In *Conservation biology: An evolutionary-ecological perspective*, ed. M. E. Soule and B. A. Wilcox, 151–69. Sunderland, Mass.: Sinauer Assoc.

Templeton, A. R. 1979. The unit of selection in *Drosophila mercatorum*. II. Genetic revolutions and the origin of coadapted genomes in parthenogenetic strains. *Genetics* 92:1283–93.

Templeton, A. R. 1986. Coadaptation and outbreeding depression. In *Conservation biology: The science of scarcity and diversity.* ed. M. E. Soule, 19–34. Sunderland, Mass.: Sinauer Assoc.

Templeton, A. R., and B. Read. 1983. The elimination of inbreeding depression in a captive herd of Speke's gazelle. In *Genetics and conservation*. ed. C. M. Schonewald-Cox, S. M. Chambers, B. MacBryde, and L. Thomas, 241–61. Menlo Park, Calif.: Benjamin-Cummings.

———. 1984. Factors eliminating inbreeding depression in a captive herd of Speke's gazelle. *Zoo Biol.* 3:177–99.

Templeton, A. R., et al. 1986. Local adaptation, coadaptation and population boundaries. *Zoo Biol.* 5:115–25.

Whitehead, G. K. 1978. *Threatened deer*. Morges, Switzerland: IUCN Publ.

Wildt, D. E., et al. 1982. Influence of inbreeding on reproductive performance, ejaculate quality and testicular volume in the dog. *Theriogen.* 17:445–51.

Wright, S. 1922. The effects of inbreeding and crossbreeding on guinea pigs. I. Decline in vigor. *Bull. U.S. Dept. Agriculture* 1093:37–63.

———. 1977. *Evolution and the genetics of populations*, Vol. 3. Chicago, Ill.: Univ. of Chicago Press.

Young, C. L., et al. 1969. Inbreeding investigations with diary cattle in the north central region of the United States. *Minn. Univ. Agricul. Exper. Stat. Tech. Bull.* 266:3–15.

6

RICHARD B. HARRIS, TIM W. CLARK,

AND MARK L. SHAFFER

Extinction Probabilities for Isolated Black-Footed Ferret Populations

Black-footed ferrets, once widespread and relatively abundant, are to-day critically endangered. Surviving populations are likely to be small isolates that are vulnerable to chance extinctions (Shaffer 1981) arising from demographic, environmental, or genetic factors. Restoration of the species to healthy numbers and distribution will require not only protection, but also knowledge of the probability of chance extinctions for isolated ferret populations of various sizes. With knowledge of extinction probabilities, recovery planning can be more solidly founded.

The use of stochastic simulation models to assess the viability of small populations was initially investigated by Shaffer (1978, 1981). The process is roughly analogous to a stress test in industrial quality control: we apply appropriate stochastic fluctuations to simulated populations in an attempt to determine the population size at which failures (extinctions) occur with acceptably low probability. In this case, we apply computer simulation to the problem of long-term management and reserve size by estimating probabilities of small ferret populations going extinct from demographic and environmental fluctuations. We define extinction as the absence of 1 of the 2 sexes. (We do not treat genetic influences on population viability because this issue has been treated by Groves and Clark [1986] and Lacy and Clark [chap. 7, this volume]. We also assume

Support for this work was provided by the U.S. Fish and Wildlife Service, the Montana Department of Fish, Wildlife and Parks, the Wildlife Preservation Trust International, and the New York Zoological Society-Wildlife Conservation International. Additional support came from the Montana Cooperative Wildlife Research Unit, the University of Montana Computing Center, and Biota Research and Consulting, Inc. Computer simulations were performed on the DEC-20 of the Digital Equipment Corporation at the University of Montana. We thank L. Ginzburg, Department of Ecology and Evolutionary Biology, State University of New York, Stony Brook, for assistance in adapting the simulation program. We also appreciate the assistance of S. Forrest, R. Oakleaf, and L. Metzgar.

that systematic extinction pressures [for example, habitat loss] have been controlled.)

Data for our hypotheses of ferret population dynamics and the resulting simulation model came from South Dakota (Hillman 1968; Henderson et al. 1969; Hillman and Linder 1973) and Meeteetse, Wyoming (Biggins et al. 1985, 1986; Clark 1986a, 1986b; Clark et al. 1986a; Forrest et al. 1985a, unpubl.). In the South Dakota study, about 90 different ferrets were seen, including 11 litters, before the population went extinct.

The dramatic population decline at Meeteetse in 1985 put the survival of the species in doubt, and short-term management to bolster the prospects of the few remaining ferrets became the first priority. It was also an appropriate time, however, to draw some lessons from our knowledge of ferret population dynamics, with the intent of developing long-term conservation strategies.

Ferret Population Dynamics

We view ferret populations as being regulated primarily through a land tenure system in which high population densities prevent juveniles from establishing themselves and thereby joining the breeding population (Richardson et al. 1987; Forrest et al. 1988). In pristine times, juvenile ferrets most likely emigrated from saturated populations and may have found vacant space in nearby prairie dog colonies. Because we focus here on the dynamics of a single, isolated ferret habitat patch, emigration can be considered functionally equivalent to death.

Data from South Dakota and Wyoming suggest that changes in natality rates play little, if any, role in population regulation. Once a female has established herself within a home range, she is likely to breed and to produce an average-sized litter regardless of environmental conditions (including ferret density), except under extreme conditions. We therefore used density-dependent survival rates and density-independent natality rates in our model. Environmental variation influenced survival rates, but not natality rates in the model.

Under equilibrium conditions, ferret populations are characterized by large seasonal fluctuations from high over-winter mortality and a relatively large cohort of juveniles (Forrest et al., 1988). Populations are at their low point in early summer (prior to parturition), and chance extinctions are most likely during this period. We view ferrets as divisible into 4 classes for purposes of survival estimates: adult females, adult males, juvenile females, and juvenile males (juveniles being young of the year). For breeding rates, we consider the additional category of year-ling females.

Meeteetse Population Data

Female ferrets first breed and produce litters as yearlings (Forrest et al., unpubl.), although we suspect that yearlings may occasionally fail to breed. Forrest et al. (1988) documented 2 cases of females over 2 years old without litters (compared with 66 adult females with litters). Deter-mining the proportion of adult females made up by these 2 cases is clouded by the fact that the sex of 14 adult ferrets without litters could not be accurately determined in the field. If all 14 were males, the breeding rate of adult females was 97% (66/68). If all were females, 16 of 82 adult females failed to produce litters, and the breeding rate was 80%. We assume the majority of unclassified adults were males.

Litter sizes varied from 1 to 5 during 1982–85 and averaged 3.32 (SD = 0.89). In 1985, however, juveniles may have been exposed to dis-temper before appearing above ground and being counted; therefore, slightly smaller 1985 litter sizes probably reflected above-normal neo-natal mortality. In addition, the potential confounding effects of the epizootic among prairie dogs (*C. leucurus*) on ferret litter sizes is un-known. Omitting 1985 data, average neonatal litter size was similar to the average litter size of 3.4 observed in South Dakota (Hillman and Clark 1980). We view variability in litter size as being largely indepen-dent of environmental factors, based on the similarity of mean litter sizes among all years in Meeteetse and of those in South Dakota, where ferret densities were much lower.

Forrest et al. (1988) documented disappearance rates at Meeteetse, but did not estimate mortality rates directly. If used as estimators for mortality, disappearance rates are biased because survival is docu-mented only when a marked animal is recaptured, but capture efficien-cies are usually less than 100%. Thus, an animal can disappear because (1) it dies, or (2) it is not recaptured. One correction for this bias is to estimate the proportion of the population that typically is not recap-tured and to increase survival rates from disappearance rates appropri-ately. For example, only 3 of 9 adult ferrets marked in 1983 were found in 1984, yielding a disappearance rate of 67% (and apparent survival of 33%). However, 29 different adult ferrets were captured during the 1984 trapping period for which spotlighting yielded a population esti-mate of 43 adults. Thus, about 14 ferrets (roughly one-third of the adult population) were not captured in 1984, suggesting an adult survival rate about one-third higher, or 44%.

Sample sizes in Forrest et al. (1988) are sufficiently small, however, that chance sampling errors have large consequences, cautioning against an overly rigid reliance on their specific values. Furthermore, while disappearance rates in Forrest et al. (1988) are not sex-specific, other data in their study indicate that male survival was lower than female survival. Adult sex ratios were biased toward females (roughly 2.2 adult females per adult male), while juvenile sex ratios, although not statistically different from 1 : 1, favored males slightly. Adult males typically occupied large home ranges than did adult females and were apparently at greater risk of predation (Richardson et al. 1987; Forrest et al. 1988). Juvenile males dispersed greater distances than juvenile females (Forrest et al. 1985b) and were more likely to leave the colony entirely.

Model Structure and Rates

The objective of the stochastic model was to simulate ferret population dynamics under hypothetical pristine conditions and to project them onto habitat isolates of different sizes. We considered survivorship and natality data from the 1985 season at Meeteetse atypical of equilibrium conditions because of the distemper epidemic. We therefore treated the 1985 survival rates as those prevailing under extremely poor environmental conditions but modeled sufficient variability in environmental conditions that such rates occasionally occurred.

The strongest data from Forrest et al. (1988) included distribution of litter sizes, age and sex ratios of the standing population, and population sizes from 1983 to 1985. Data on age-specific breeding rates and survival were based on small samples, and we had less confidence in their accuracy and precision. Rates applied in the model are summarized in Table 6.1.

We assumed throughout a 95% breeding rate for females over 2 years old (Table 6.1). We assigned yearlings an 85% breeding rate, based largely on data for American martens (*Martes americana*) suggesting that yearlings are less effective breeders than older females (Strickland et al. 1983; Hass and Wright, unpubl. data, 1986). Breeding of each female was determined by comparing her breeding rate (either 0.85 or 0.95) with a uniformly distributed random (0,1) number.

Litter sizes of each bred female were computed as the nearest integer to a normal random variate with a mean of 3.38 and a standard deviation of 0.86, truncated at litter sizes of 1 and 5 (Table 6.1). We allowed for no variation in mean litter size among years. Our model thus incorporated

Table 6.1. Parameters Used in Simulation Model of Ferret Population Dynamics

	Age Class		
Parameter	Juvenile	Yearling	2+
Breeding rate	0.00	0.85	0.95
Mean litter size	—	3.38	3.38
Standard deviation	—	0.86	0.86
Female survival at equilibrium	0.28	0.60	0.60
Male survival at equilibrium	0.15	0.50	0.50
Proportion of initial age structures:			
Females	0.34	0.10	0.12
Males	0.34	0.06	0.04

only demographic variability in natality, in contrast to the Shaffer model (1978, 1983), which incorporated both demographic and environmental variability in natality.

We devised density-dependent survivorship schedules by assuming the Meeteetse population of 1984 to be at or near carrying capacity (K), and by first estimating survival rates at K. We manipulated survivorship rates of the 4 age classes until: (1) adult sex ratios and juvenile-to-adult ratios approximated those observed by Forrest et al. (1988); (2) survival rates, when applied to the natality rates described above, generated population trajectories similar to those observed during 1982–84; and (3) relative survival rates among the 4 sex and age classes followed the pattern described by Forrest et al. (1988). The resulting survival rates at K (Table 6.1) constituted a plausible reconciliation of the disappearance rates, sex and age ratios, and population trends reported by Forrest et al. (1988).

We then estimated survival rates for lower and higher densities than observed at Meeteetse, based on our views of ferret population regulation summarized previously (Fig. 6.1). At population levels greater than K, juvenile survival drops precipitously, while adult survival remains relatively constant. At low densities, all age groups have much higher survival rates, although juveniles still suffer greater mortality than adults, and males suffer greater mortality than females. In the model, survival of each individual was calculated separately, by comparing its probability of surviving with a random, uniform (0,1) number.

We also added a component of environmental variability to survival

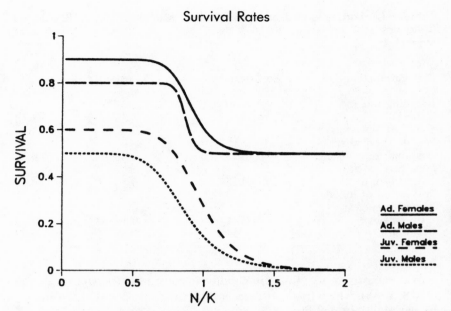

Figure 6.1. Modeled survival rates of adult females, adult males, juvenile females, and juvenile males, as functions of the relative density of the population (*N/K*). (See text for justification of the shapes.)

rates at all densities by treating each as the mean of a normal distribution with a specified variance. The standard deviations used in modeling were constant proportions of the mortality rates, so the model input actually consisted of coefficients of variation (CVs) of mortality rates. Ferret populations have not been studied long enough to estimate the variation in yearly survival rates. Consequently, we bounded the likely values by running all simulations with mortality rate CVs of 0, 10, and 25%. The frequency with which extremely high or low mortality rates occurred in model runs can be calculated by applying standard z values to the standard deviations of mortality (expected rate multiplied by CV). With 10% CV of mortality rates, approximately 68% of expected juvenile male mortality rates fell between 0.765 and 0.935; for juvenile females the comparable range was 0.648 to 0.792. With 25% CV of mortality rates, approximately 16% [0.5 × (1 − 0.68)] of runs had expected mortality rates greater than 1.0, 0.90, 0.625, and 0.5 for juvenile males, juvenile females, adult males, and adult females respectively.

Sex and age ratios of the standing population at Meeteetse from 1982 to 1984 were relatively constant. We used initial populations in the model that, scaled appropriately, approximated these sex and age ratios

(Table 6.1). In exploratory model runs, we tried initial population structures that reflected the most extreme ratios reported from Meeteetse and found that turnover in modeled ferret populations was rapid enough to make the initial age structure irrelevant within a few years. Because our primary focus was on ferret management over periods of 20 to 100 years, we felt justified in using only 1 initial age structure per population size.

The resulting simulation model portrayed ferret populations under hypothetical pristine conditions. Modeled populations for each year fluctuated around K, from high numbers just after the birth pulse to low numbers following winter mortality. Variability resulted from demographic stochasticity in survival, proportion of females breeding, litter size, and sex ratio of newborns. Environmental variability was added by treating expected mortality rates as random variables from normal distributions. All runs were of 100-year duration, and we tallied extinctions that occurred at 10-year intervals. For each population size we ran the model 500 times (Harris et al. 1987).

Because we were unsure of the precise nature of the density-dependent response in survival rates, we also ran the model with weaker and stronger responses to low density. In both cases, survival rates at equilibrium were kept at those shown in Table 6.1, thus maintaining expected population size at their equilibria. In these alternative runs, however, we altered all survival rates by adding or subtracting 0.20 whenever the population fell below one-half its equilibrium value ($N/K < 0.5$). In this way, expected population size in the absence of any variation was equivalent to the normal runs, but the strength of the compensatory response of a population to chance excursions into very low densities was varied.

Results

Modeled populations went extinct at relatively constant rates through time, resulting in exponential declines in persistence probability (Fig. 6.2). The smallest populations showed a slight tendency for persistence probability to be higher during the first 10 years than during the following 90, probably because initial population structures were slightly more favorable for persistence than the stable age distributions that developed later.

In all sets of simulation runs, the relation between persistence probability and population size was nonlinear. Persistence probability increased slowly at very low equilibrium populations, then rapidly

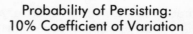

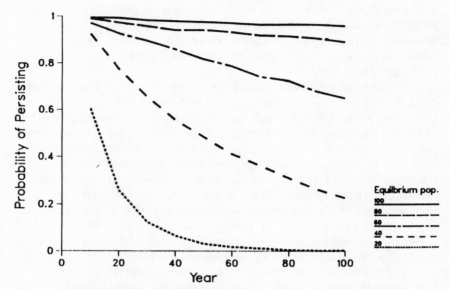

Figure 6.2. Probability of population persistence through time for populations with carrying capacities of 20, 40, 60, 80, and 100 animals. All runs included 10% coefficient of variation (CV) of mortality rates. Results are based on 500 independent simulation runs.

through a range of population sizes, with an asymptote at probabilities greater than 80% (Fig. 6.3a).

Population sizes that produced 95% persistence varied depending on the degree of environmental variability modeled. In runs with no environmental variability (all variation due solely to chance demographic events), an equilibrium population size of 70 was the smallest that had at least a 95% chance of lasting 50 years, and a population of 80 the smallest having a 95% chance of lasting 100 years (Fig. 6.3A). Adding 10% CV to mortality rates increased these equilibrium population sizes to 90 and 100, respectively (Fig. 6.3B). Adding 25% CV to mortality rates further increased these equilibrium population sizes to 120 and 150, respectively (Fig. 6.3C).

Persistence probabilities with 0 and 10% CV in mortality rates were fairly similar (Figs. 6.3A and 6.3B), suggesting that most extinctions at small population sizes occurred primarily due to chance demographic events and that environmental variability was a secondary influence. Furthermore, the magnitude of compensatory survival at low densities had only a minor influence on extinction probabilities (Table 6.2 and Fig. 6.3D). Extinction probabilities were little changed when we altered survival rates at low densities, because few extinctions were preceded by random drifts into numbers below $0.5K$. Extinctions were more often preceded by population levels greater than K, at which our hypothesized juvenile survival rates fell to near 0. When the number of remaining adults was small, chance fluctuations in adult mortality then made these populations vulnerable to collapse. Males were usually less numerous than females due to their higher mortality rates, and most extinctions were caused by the complete loss of males in the population.

Our synthesis of black-footed ferret dynamics suggests that populations averaging less than 100 are vulnerable to chance extinctions largely because of the pronounced seasonal fluctuations in numbers that characterize the species. Because such a large proportion of the summer population consists of nonbreeding juveniles (68% at Meeteetse during 1982–84), the burden of reproduction rests on a relatively small number of breeding females and an even smaller number of breeding males. Our analysis of age structures suggests that with a total summer population size of 50 animals, only 2 are expected to be males over 2 years old. (Table 6.1). Chance deviations in the survival of individual males, or in the sex of offspring, can easily lead to the complete absence of breeding-age males in the population. Additionally, natural mortality rates for the juvenile age class are high enough that total recruitment failures are not uncommon.

The magnitude of environmentally caused variability in ferret population dynamics (the variation among years in expected birth and death rates) is unknown. Our model containing no environmental variability is clearly unrealistic, however, because survival rates during the 1985 de-

Table 6.2. Persistence Probability over 100 Years of Alternative Model Assumptions

| | Equilibrium Population Size | | | |
Assumptions	30	50	70	100
10% CV, mortality	.002	.428	.788	.958
+20% mortality when $N/K < 0.5$	.012	.298	.730	.938
−20% mortality when $N/K < 0.5$	.050	.462	.802	.952

Note: The first line shows results from Fig. 6.3b. The second and third line are results using models with 10% CV in mortality rates, and in which mortality of all age classes was increased or decreased by 20% whenever the population fell below $0.5K$.

A

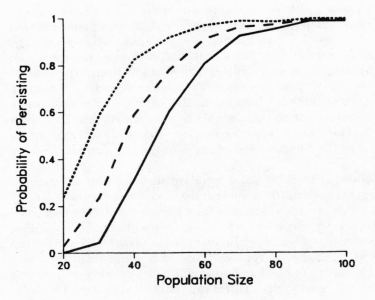

B

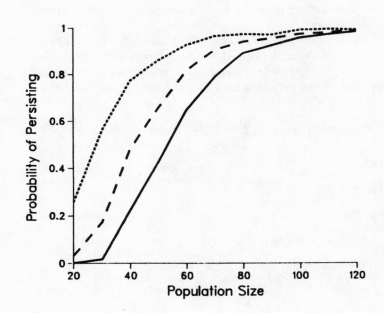

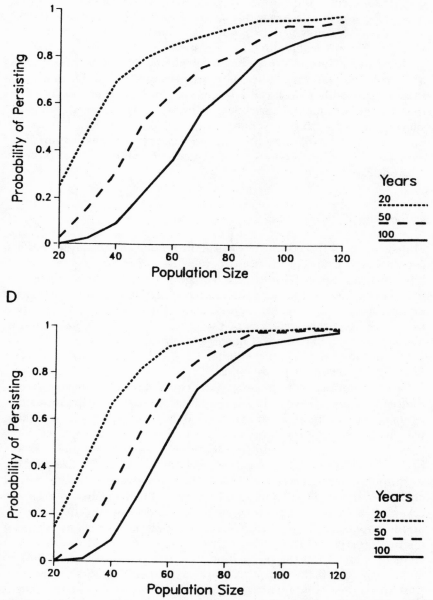

Figure 6.3. Probability of population persistence as a function of population size for 20, 50, and 100 years: (A) No environmental variability modeled. (B) Environmental variability modeled by adding 10% CV to all mortality rates. (C) Environmental variability modeled by adding 25% CV to all mortality rates. (D) Same as B, but with 20% added to all mortality rates whenever the population level fell below 0.5*K*.

cline at Meeteetse were lower than allowed by this model. The model containing mortality rates with a 25% CV probably represents a pessimistic case. The intermediate model, with mortality rates varying with a CV of 10%, is probably the most realistic.

However, differences in extinction probabilities under these 3 levels of environmental variability were surprisingly small. Demographic stochasticity evidently played a larger role in these chance extinctions than did environmental variability at the levels we modeled. Similarly, varying the strength of responses of modeled populations to very low densities made surprisingly little difference in resultant extinction probabilities, because extinctions were as likely to follow high population levels as low.

Management Implications

Management of ferrets must be based on a multitude of considerations, avoidance of chance extinctions being just one. Our results suggest, however, that ferret populations averaging 20 to 120 will likely go extinct periodically within 100 years, even under total protection. Thus, maintenance of multiple, independent populations provides the best insurance against overall loss of the species from chance extinctions. These small populations will need close monitoring and in periods of severe decline or extirpation will need augmentation of surplus ferrets from larger, healthier populations.

Given that multiple populations are required for long-term recovery, the relationships depicted in Figure 6.3 can be used to help guide managers interested in minimizing the risk of chance extinctions in making trade-offs among a few large reserves and many small reserves. Very small habitat patches will support ferret populations with unacceptably high probabilities of chance extinctions. Increasing patch size (and, by assumption, ferret-carrying capacity) will greatly reduce the risk of chance extinctions until a point is reached at which further increases in carrying capacity yield proportionally less and less protection. For example, suppose habitat patch A is large enough to support an average of about 40 ferrets. The doubling of area A will greatly reduce the risk of chance extinction of the population over 100 years, from over 80% to about 10% (Fig. 6.3b). Suppose, however, that habitat patch B can support an average of 80 ferrets. In this case, the addition of an equivalent amount of habitat would better reduce the risk of chance extinctions if used to start a new, independent population or to augment a smaller patch, such as A.

Clark et al. (1987) have identified 8 prairie dog complexes in Montana as potential ferret transplant sites. The smallest complex is 577 hectares and the largest is 3,147 hectares. Assuming that these black-tailed prairie dog colonies (*C. ludovicianus*) can support ferrets at similar mean densities to those observed in South Dakota (1 ferret litter/36 hectares [Hillman and Linder 1973]), these complexes may support from 16 to 87 adult females each year. Densities of adult females dependent on white-tailed prairie dogs (*C. leucurus*) at Meeteetse were 1/57 hectares (Forrest et al. 1985). The Montana complexes could support from 10 to 55 adult females at this lower density. Extrapolating from the age structures observed at Meeteetse (Forrest et al. 1988), average ferret numbers (including juveniles) on the Montana complexes could thus be expected to vary from 45 to 395. Based on our modeling results, a population on the largest complex would be relatively immune from the risk of chance extinction over a 100-year time frame. The more numerous smaller colonies would be expected to go extinct occasionally, but occasional reintroductions from the larger colonies could serve to keep the entire metapopulation (Levins 1970) viable indefinitely.

References

Biggins, D. E., et al. 1985. Movements and habitat relations of radio-tagged black-footed ferrets. In *Black-footed ferret workshop proc.*, ed. S. Anderson and D. Inkley, 11.7–11.17. Cheyenne, Wyo.: Wyoming Game and Fish Publ.

Biggins, D. E., et al. 1986. Activity of radio-tagged black-footed ferrets. *Great Basin Nat. Memoirs* 8:135–40.

Clark, T. W. 1986a. Technical introduction. *Great Basin Nat. Memoirs* 8:8–10.

Clark, T. W. 1986b. Some guidelines for management of the black-footed ferret. *Great Basin Nat. Memoirs* 8:160–68.

Clark, T. W., J. Grensten, M. Gorges, R. Crete, and J. Gill. 1987. Analysis of black-footed ferret translocation sites in Montana. *Prairie Naturalist* 19:43–56.

Clark, T. W., et al. 1986a. Descriptive ethology and activity patterns of black-footed ferrets. *Great Basin Nat. Memoirs* 8:115–34.

Clark, T. W., et al. 1986b. Description and history of the Meeteetse black-footed ferret environment. *Great Basin Nat. Memoirs* 8:72–84.

Forrest, S. C., et al. 1985a. Black-footed ferret habitat: Some management and reintroduction considerations. *Wyoming Bur. Land Mgmt. Wildl. Tech. Bull.* no. 2, Cheyenne, Wyo.

Forrest, S. C., et al. 1985b. Life history characteristics of the genus *Mustela*, with special reference to the black-footed ferret, *Mustela nigripes*. In *Black-footed ferret workshop proc.*, ed. S. Anderson and D. Inkley, 23.1–23.14. Cheyenne, Wyo.: Wyoming Game and Fish Publ.

Forrest, S. C., et al. 1988. Black-footed ferret (*Mustela nigripes*) population attributes, Meeteetse, Wyoming, 1981–1985. *J. Mammal.* 69:261–73.

Groves, C. R., and T. W. Clark. 1986. Determining minimum population size for recovery of the black-footed ferret. *Great Basin Nat. Memoirs* 8:150–59.

Harris, R. B., L. A. Maguire, and M. L. Shaffer. 1987. Sample sizes for minimum viable population estimation. *Conservation Biology* 1:72–76.

Henderson, F. R., et al. 1969. The black-footed ferret in South Dakota. *South Dakota Dept. Game, Fish and Parks Tech. Bull.* no. 4.

Hillman, C. N. 1968. Field observations of black-footed ferrets in South Dakota. *Trans. North Amer. Wildl. Nat. Resour. Conf.* 33:433–43.

Hillman, C. N., and T. W. Clark. 1980. *Mustela nigripes. Mamm. Species* no. 126.

Hillman, C. N., and R. L. Linder. 1973. The black-footed ferret. In *Proc. black-footed ferret and prairie dog workshop,* ed. R. L. Linder and C. N. Hillman, 10–23. Brookings, S. Dak.: South Dakota State Univ.

Levins, R. 1970. Extinction. In *Some mathematical questions in biology,* ed. M. Gerstenhaber, Vol. 2, 77–107. Providence, R.I.: Amer. Math. Soc.

Richardson, L., et al. 1987. Winter ecology of the black-footed ferrets (*Mustela nigripes*) at Meeteetse, Wyoming. *Am. Midl. Nat.* 117:225–39.

Shaffer, M. L. 1978. Determining minimum viable population sizes: A case study of the grizzly bear (*Ursus arctos* L.). Ph.D. diss., Duke Univ., Durham, N.C.

Shaffer, M. L. 1981. Minimum population sizes for species conservation. *Bioscience* 31:131–34.

Shaffer, M. L. 1983. Determining minimum viable population sizes for the grizzly bear. *International Conference on Bear Research and Management* 5:133–39.

Strickland, M. A., et al. 1983. Marten (*Martes americana*). In *Wild mammals of North America,* ed. J. A. Chapman and G. A. Feldhammer, 599–612. Baltimore, Md.: Johns Hopkins Univ. Press.

7

R O B E R T   C .   L A C Y   A N D   T I M   W .   C L A R K

Genetic Variability in Black-Footed Ferret Populations: Past, Present, and Future

This chapter estimates effects of historical and recent declines in black-footed ferret numbers on genetic variability within the species. It also examines genetic variability within the surviving ferret population under several possible scenarios of demographic recovery. Understanding these genetic considerations is essential to ferret recovery planning and management.

Genetic Consequences of Rarity

When a sexually reproducing population is reduced to few numbers, genetic changes occur that can lead to further decline and make eventual demographic recovery less likely. The random sampling of genes at each new generation changes frequencies of allelic variants (genetic drift), even in the absence of the other evolutionary forces of selection, mutation, and migration. In any sexual population lacking mechanisms to restore genetic variation, genetic variability will slowly be depleted, as alleles occasionally drift to fixation or extinction.

Random drift is much more rapid in small populations than in large, because each generation is a less representative sample of the previous gene pool. The loss of heterozygosity (the most common measure of genetic variation, proportional to the variance in allele frequencies) expected in one generation due to genetic drift acting on selectively neutral traits is given by the equation

The Chicago Zoological Society has supported the research of both authors. The New York Zoological Society-Wildlife Conservation International, Wildlife Preservation Trust International, National Geographic Society, and World Wildlife Fund-U.S. provided support for work on the black-footed ferret to Tim W. Clark.

$$H_t = H_{t-1} [1 - 1 / (2N_e)] \tag{7.1}$$

in which H_t and H_{t-1} are the heterozygosities at generations t and $t - 1$, and N_e is the effective population size (Crow and Kimura 1970). The effective population size (the inbreeding effective number) is the size of an idealized random mating population with an equal sex ratio and a Poisson distribution of family sizes that would lose heterozygosity at the same rate as does the observed population. An unequal sex ratio, assortative mating, or greater than Poisson variance in family sizes will cause heterozygosity to be lost at a greater rate and the effective population size to be less than the census population size. Disassortative mating or less than Poisson variance in family sizes will slow genetic drift (because the offspring generation is a more even representation of the parental gene pool) and cause the effective population size to exceed the census number.

Over a number of generations (or a fractional generation in a population with continuous generations), the cumulative loss of heterozygosity is approximated by

$$H_t = H_0 [1 - 1 / (2N_e)]^t \tag{7.2}$$

in which H_t and H_0 are the heterozygosities at generations t and 0, and the effective population size, N_e, is assumed to remain constant.

A severe decline in the number of animals in a population, resulting in a very few survivors for one or more generations, is referred to as a population bottleneck. The effect of population decline on genetic variability in a population is highly dependent upon the duration of the bottleneck, little variation being lost if the population recovers within a few generations (Nei et al. 1975). Ralls et al. (1983) examined the expected loss of heterozygosity during the decline and subsequent recovery of California sea otters (*Enhydra lutris*), estimating effective population size under a variety of demographic assumptions. They determined that, because of the short duration of the bottleneck, the sea otter population should retain a large fraction of its initial heterozygosity. Their prediction was supported by the finding of 6% heterozygosity in the sea otter population (Lidicker and McCollum 1981), although no rare alleles were found. Northern elephant seals (*Mirounga angustirostris*) experienced a similar bottleneck, but have been found to have little or no genetic variability (Bonnell and Selander 1974). O'Brien et al. (1983, 1985) found no genetic variability in South African cheetahs (*Acinonyx jubatus*) and speculated that the cheetah has experienced a severe population bottleneck in the past. The effect that a sudden decline in heterozygosity has on a population, and thus the probability of recovery from a severe population crash, probably depends on the genetic variation present prior to the decline. No natural population of a vertebrate has been examined genetically before, during, and after a population bottleneck, and it is possible that postbottleneck measures of genetic variability reflect initial genetic conditions as much as effects of the population decline.

Loss of heterozygosity commonly leads to lower fecundity and viability (inbreeding depression) and can therefore cause the population to decline still further. Increasingly rapid losses of heterozygosity and further inbreeding-induced population decline then create a positive feedback that may drive a population to extinction. Even if the loss of heterozygosity does not significantly reduce fitness of individuals, perhaps because the population became adapted to a state of lower heterozygosity (as would be the case if selection eliminated those alleles, such as deleterious recessives, causing inbreeding depression), the population as a whole can lose the evolutionary flexibility conferred by genetic diversity. Without genetic variation between individuals on which natural selection can act, a population cannot adapt to changing environmental conditions. For example, a severe response by a captive cheetah population to infection by feline infectious peritonitis has been interpreted as evidence for species vulnerability in the absence of genetic variation (O'Brien et al. 1985). In a small, localized population, selection for survival in a specific habitat may accelerate the loss of genetic diversity, making it even less likely that a population would ever expand into more diverse habitats.

Natural selection can maintain genetic variation if it favors heterozygotes or if different genotypes are favored in different environments. Immigration into a population from genetically distinct populations can restore variation lost by drift or selection, though obviously only if divergent populations of the species exist and migration between the populations is possible.

Ultimately, new variation must be introduced by mutation. Mutation, typically at rates of 10^{-6} to 10^{-5} per locus per generation, is too infrequent to be important to very small populations over short time scales (Lacy 1987). Mutation provides a substantial relative increase in heterozygosity only after a population has become almost devoid of genetic variation. Moreover, most genetic variants lost from a formerly large population by genetic drift during a population decline would have been selectively neutral or even advantageous in some environments. Most new mutations would be deleterious and rapidly eliminated by natural selection, or neutral or nearly neutral and of no adaptive consequence. The restoration of genetic variants that provide short-term

increases in fitness (by heterozygote advantage) or long-term evolutionary flexibility (by adaptation to alternative environments) might be expected to be several orders of magnitude slower than the mutation rate
itself.

Investigation of the genetic effects of population decline in ferrets
(estimating the level of heterozygosity in the ancestral populations, the
subsequent loss of variability, and the expected genetic fate of the species) requires estimates of effective population sizes for use in Equation
7.1 or 7.2. To calculate effective population sizes, demographic data are
needed on:

1. breeding adult population size each generation, or average population
   size with the variance across generations;
2. average generation length, calculated from fecundity and mortality
   schedules;
3. sex ratio of the breeding population;
4. average litter size and variance in litter size; and
5. distribution of juvenile mortality among families.

The Meeteetse Population

If efforts to save the black-footed ferret are successful, including breeding ferrets in captivity and restoring viable populations in the wild,
recovery plans will have to consider the preservation or restoration of
genetic variability, as well as population growth and stability. The current genetic status of the Meeteetse ferrets is unknown, but Kilpatrick et
al. (1986) found no variation among 22 ferrets in an electrophoretic
analysis of 3 salivary proteins. O'Brien et al. (chap. 3, this volume) studied genetic variation in the Meeteetse ferrets by electrophoresis of blood
and tissue proteins.

The history of the Meeteetse prairie dog and ferret complex, focusing
on its size and distribution and destruction via prairie dog poisoning
from about 1900 to the present, is given by Clark et al. (1986). The
Meeteetse ferret population historically (ca. 1930) occupied a prairie
dog complex of about 8,400 hectares and, assuming 1 adult ferret per 57
hectares as in the Meeteetse area, would have supported about 147 adult
ferrets (Clark et al. 1986). Details are sketchy, so this historical picture is
partially speculative and represents only a broad outline. Sporadic prairie dog poisoning began in the 1880s and well-organized control programs (1935 to 1940) probably reduced the ferret population significantly. Scattered, localized poisoning occurred over the next 3 decades.

Subsequently, in 1981, about 3,000 hectares of prairie dog colonies remained and supported about 43 adult ferrets at peak numbers in 1984 (Forrest et al. 1985a, 1988).

We estimated the size of the Meeteetse ferret population in 1935 and thereafter from the history of prairie dog colonies in the area. Two sets of estimates were made. In the "best-case" scenario, shown by the upper line in Figure 7.1, we assumed that poisoning programs of the 1930s reduced the population from 147 to 50 adults, with a linear decline from 1935 to 1940, and that the population remained relatively stable at that level through 1981. This best-case scenario assumes that the Meeteetse population had remained at an average size as great as or greater than has been observed since 1981. In our "worst-case" scenario, shown by the lower line in Figure 7.1, we assumed that the Meeteetse ferret population was reduced to only 10 animals in 1940 and that it increased to 15 animals by 1945 and remained at that low level through 1981. In this scenario it is assumed that the Meeteetse population (30 to 43 adult

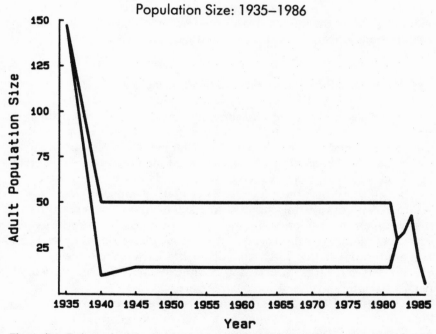

Figure 7.1. Estimated adult population sizes for black-footed ferrets (BFF) in the Meeteetse, Wyoming, area from 1935 to 1986. An optimistic ("best-case") scenario and a pessimistic ("worst-case") scenario are presented as approximate boundaries on the probable size of the Meeteetse population.

ferrets in 1982 to 1984) was recovering in numbers between its discovery in 1981 and the 1985–86 decline.

In both the best-case and worst-case scenarios, we based 1982 through 1985 population estimates (30, 34, 43, and 20 adult ferrets) on population data reported by Forrest et al. (1988) and Clark et al. (1986). Spotlight counts were conducted over part of the Meeteetse area in 1982. By assuming that the surveyed area contained the same proportion of the total ferret population that was observed during extensive surveys of 1984 and 1985, and that two-thirds of the ferrets counted were juveniles, it was estimated that the Meeteetse population consisted of 30 adults in 1982. In 1983, 88 ferrets were seen in extensive spotlight surveys, but mark-recapture estimates yielded 113 animals. Averaging these 2 types of estimates, and assuming two-thirds of the ferrets to be juveniles, it was estimated that the population consisted of 34 adult ferrets in 1983. Thorough spotlight surveys and mark-recapture studies gave estimates of about 43 adult ferrets in 1984 and 20 in 1985. Based on the subsequent captures of 2 adult females with litters and 2 adult males, we very tentatively estimate that perhaps 6 adult ferrets persisted in the wild just prior to the fall 1986 captures, and we use this number for the 1986 population size in both our scenarios.

Population Structure and Demographic Parameters

About 1920, when campaigns to eliminate prairie dogs from ranchlands were moving into high gear in western North America, prairie dog colonies covered an estimated 40.5 million hectares of western rangeland (Anderson et al., 1986). At an estimated density of 1 ferret per 57 hectares of prairie dog colony, as found at Meeteetse (Forrest et al. 1985a), there may once have been as many as 800,000 ferrets. We assumed, with prairie dog complexes of 500 to 10,000 hectares or more per complex in close enough proximity to support an interbreeding ferret population, local ferret populations may have consisted of 10 to 100 adult animals (see Flath and Clark 1986). Scattered prairie dogs existed between large prairie dog complexes, but few ferrets would persist for long outside of the complex areas.

Forrest et al. (1985a, 1988) reported ferrets disperse an average of 2.5 km between prairie dog colonies, with a maximum recorded intercolony movement of 5.7 km. This is considerably less than typical distances between many remaining complexes of prairie dog colonies today, and between 1982 and 1984 no ferrets were observed to disperse from the primary Meeteetse complex to another complex 38 km away (Forrest et al. 1985a). The patchy nature of the prairie dog colonies within the Meeteetse area complex is shown in Figure 7.2. Ferret populations thus probably consisted historically of small demes, each occupying a local complex of prairie dog colonies and being fairly panmictic, but connected to neighboring demes by occasional migrants.

The Meeteetse population has an adult sex ratio of approximately 1 male:2 females, with a juvenile sex ratio not statistically different from

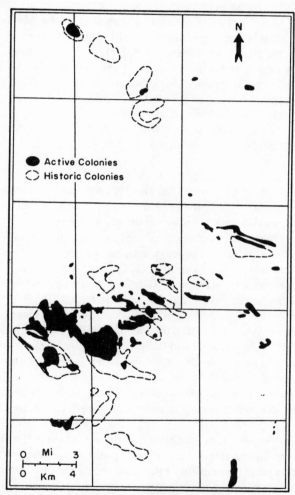

Figure 7.2. Location of active and historic ("dead") prairie dog colonies near Meeteetse, Wyoming, in 1984 (Clark et al. 1986).

1 : 1. Mean litter size is 3.3 (SD = 0.89, n = 68 litters), with a range of 1 to 5 (Forrest et al. 1988). From 53 to 86% of the ferrets disappear annually (Forrest et al. 1985a, 1988), and data on other mustelids (reviewed in Forrest et al. 1985b, 1988) suggest that mortality may be as high as 89% the first year of life and over 50% per year thereafter. Adults breed only once per year. We assume for the purposes of our calculations that the average generation time is 1.5 years, as would result from a 67% annual adult mortality. Captive ferrets could have much lower mortality and thus a longer average generation time (Richardson et al. 1986).

Given the high birth and death rates, as seen in the Meeteetse ferret population, local ferret populations would undergo considerable fluctuations from year to year, although the Meeteetse area may have unusually high raptor densities (Phillips, pers. comm., 1985). From 1982 through 1985, the Meeteetse population averaged 31.75 adult ferrets, with a standard deviation of 9.535 (CV = 30.03%) and a range of 20 to 43 (Clark et al. 1986; Forrest et al. 1988).

Based on this brief review of the Meeteetse ferret population and habitat characteristics, we assumed the following characteristics in our genetic analyses:

1. Population sizes between 1935 and 1986 were between the best-case and worst-case scenarios presented in Figure 7.1. The coefficient of variation (SD/N) in population size across generations is 0.3003.
2. Ferrets breed at 1 year of age (if they survive that long), a few survive to reproduce in subsequent years, and average generation time is 1.5 years.
3. The sex ratio of breeding adults is 1 male:2 females.
4. At emergence from the den (8 weeks postpartum), but prior to juvenile mortality, mean litter size is 3.3, with variance 0.79.
5. Juvenile (prereproduction) mortality falls between 2 extreme cases: random mortality of juveniles, the fate of each ferret being independent of the fates of littermates; and family mortality, entire litters dying or surviving as a unit. Juvenile mortality was set (inherent within the calculations of the model) to be of the magnitude that would reduce the postreproduction fall population to the size of the breeding population in the subsequent spring, as given in Figure 7.1.

We assume selective neutrality of genetic variation throughout our calculations in order to provide a basic point of departure for further understanding and discussion of the losses of variability likely to occur in ferrets and other endangered species occurring in small populations. Unless selection has strongly affected variation in ferrets in the recent past or does so in the near future, the assumption of neutrality should allow reasonable estimates of genetic changes resulting from the decline of ferrets.

Genetic Variability Prior to 1880

Electrophoretic surveys of other mustelid species (see Kilpatrick et al. 1986) have revealed very low levels of heterozygosity. Lidicker and Mc-Collum (1981) reported 6% heterozygosity in sea otters, but Simonsen (1982) found no variation in 21 protein-coding loci analyzed in 6 other mustelid species. A survey of electrophoretically detectable variation at 33 loci in a sample of 34 North American river otters (*Lutra canadensis*) from a large, wild population in Louisiana revealed only 1 polymorphic locus and an average heterozygosity of just 0.6% (Lacy, unpublished data). A poorly resolved locus appeared to be variable also; including it in the analysis may double the heterozygosity estimate for river otters. Kilpatrick et al. (1986) summarize electrophoretic data on 14 carnivore species, reporting a mean heterozygosity of 1.4%, lower than the mammalian mean of 3.6% reported in a compilation of 48 species by Nevo (1978).

Ferret populations may have been more isolated (patchy) than are other North American *Mustela*, and population densities and dispersal rates are probably low compared to larger carnivores (for example, canids). Ferrets may historically have been adapted to very low levels of heterozygosity, and this may have contributed to their ability to persist at low numbers over the past 50 years. Although not close taxonomically, beaver (*Castor canadensis*) may have population structures and dispersal rates analogous to those in ferrets, with beaver ponds corresponding to the prairie dog colonies supporting ferrets and drainage systems corresponding to prairie dog complexes. Hoppe et al. (1984) found 3 of 34 proteins to be variable, with an average heterozygosity of 1.0%, in a sample of 48 beaver from a large population in South Carolina that was restocked by multiple introductions more than 20 years earlier.

Based on an estimated effective local population size of 100, ferret populations would have maintained only about 0.04% heterozygosity when in mutation-drift balance, more than an order of magnitude lower than is commonly observed in carnivores. (The heterozygosity maintained by mutation-drift balance is [Crow and Kimura 1970]

$$H = 1 - 1/(4N_e m + 1) \tag{7.3}$$

in which m is the mutation rate per locus per generation [typically 10^{-6}].)

Genetic differentiation probably existed among ferret populations on isolated prairie dog complexes, though migration between complexes could have prevented such differentiation and kept within-population heterozygosity much higher than estimated above.

Genetic Variability, 1935–1986

The rate of decline of heterozygosity due to genetic drift is inversely proportional to the effective population size (Eq. 7.1). The effective population size is in turn a function of the mean and variance in census population size, sex ratio of breeding adults, and variance in family size (measured at sexual maturity). The equations below, for adjusting the effective population size for these factors, are given in Crow and Kimura (1970) and other population genetics references. Over many generations, the cumulative loss of heterozygosity can be calculated iteratively (if the population is censused each generation), or equivalently the geometric mean effective population size can be applied to each generation. Alternatively, the mean effective population size can be corrected for variance in population size by

$$N_e = N \,/\, [1 + (V_N/N^2)] \tag{7.4}$$

in which N and V_N are the mean and variance in population size. The coefficient of variation $[(V_N)^{0.5}/N]$ observed in ferrets from 1982 to 1985 is 0.3003. Thus the correction in N_e for yearly fluctuations in population size can be estimated as $.9173N$ and should be applied to those periods (prior to 1982) when annual censuses are not available and population numbers were set equal to some assumed average level.

An unequal breeding sex ratio depresses the effective population size by

$$N_e = 4N_m N_f \,/\, (N_m + N_f) \tag{7.5}$$

in which N_m and N_f are the numbers of males and females, because the genes contributed by the rarer sex contribute disproportionately to the next generation. For the $1:2$ sex ratio in ferrets, the effective population size is reduced by a factor of 0.8889 relative to a $1:1$ sex ratio.

Variance in family size affects the effective population size of species with separate sexes breeding at random according to

$$N_e = (2N - 2) \,/\, (V_k/k + k - 1) \tag{7.6}$$

in which k and V_k are the mean and variance in family size. If the population is stable ($k = 2$) and family sizes follow a Poisson distribution ($V_k/k = 1$), then $N_e = N - 1$. In ferrets, and probably many other vertebrates, litter sizes at birth show less variance than Poisson, so that the effective population may be greater than the census population size.

The variance in family sizes at breeding, however, depends critically on the nature of mortality of the litter. If juvenile mortality is clustered within litters (a litter tends to survive or die as a unit), then the variance in family size at reproductive age will be much greater than the variance of litter sizes at birth. Conversely, if mortality occurs at random with respect to sibships, variance in family size will be closer to Poisson than is variance in litter size. Crow and Morton (1955) give the equations for adjusting observed variances in litter sizes at birth for mortality of offspring prior to their reproduction: for random survival of young,

$$V_{k'} = k'[1 - (k'/k)] + (k'/k)^2 V_k \qquad (7.7)$$

for survival of entire litters,

$$V_{k'} = [k'/k]V_k + k'k(1 - (k'/k)] \qquad (7.8)$$

in which $V_{k'}$ and V_k are the variances in family size after mortality occurs and before mortality occurs, and k' and k are the mean family sizes after and before mortality. Usually mortality would be partially clustered by family, and the above equations would provide upper and lower limits on the effective population sizes. For ferrets, with $k = 3.3$ and $V_k = 0.7921$, $V_{k'}$ is 1.0788 and 3.0800 in the cases of random and family mortality when the population is stable ($k' = 2$). If the population is unstable, $k' = 2N_t / N_{t-1}$, and $V_{k'}$ can be determined each generation from Equations 7.7 and 7.8. Mortality due to epizootics, or to predation on lactating females or their litters within burrows, would affect entire litters. Because most ferret mortality occurs following offspring dispersal from the natal burrow, however, ferret mortality probably more closely approximates the random case, and variance in family size is probably less than Poisson even after juvenile mortality.

Although the separate effects of sex ratio, variance in family size, and variance in population size on the effective population size (and thus on the loss of genetic variability) have been determined, it is not known how multiple effects should be compounded to estimate accurately the rate of genetic drift within a population. When each effect is small, the decrease in effective population size and the loss of heterozygosity is approximately linearly related to sex ratio, to variance in family size, and to variance in the population size, and an iterative application of the above equations (implying multiplicative effects) is probably appropriate. Changing the order in which Equations 7.4, 7.5, and 7.6 are applied will affect the final estimate obtained, though not greatly. If demographic parameters are known to differ between males and females, corrections should be made for variances in litter size and in population

numbers to determine effective numbers of males and females separately, and then Equation 7.5 applied to determine an overall effective
population size (Crow and Kimura 1970).

Estimated effective population sizes for the Meeteetse ferrets during
the years 1935 to 1986 are shown in Figure 7.3. Adjustments were made
iteratively for the 1 : 2 sex ratio and for variance in family size, with an
assumption of either random mortality of juveniles or whole-litter mortality. Population sizes for years between estimates were calculated assuming linear declines and increases as shown in Figure 7.1. However,
effective population sizes for years prior to 1982 were adjusted by Equation 7.4, assuming that the population fluctuates annually with a coefficient of variation of 0.3003 (as calculated from 1982 to 1985 census
data). Although the estimated variance in population size is based on
census data from only 4 years, improvement of the estimated variance
would have a minor effect on the final estimates of effective population
sizes and heterozygosities.

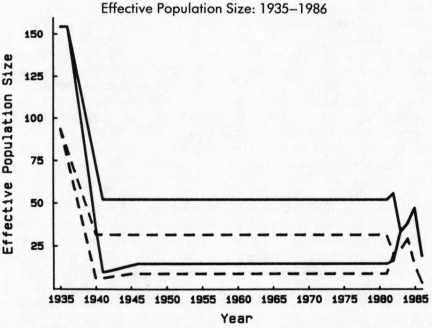

Figure 7.3. Effective population sizes of the Meeteetse ferret population, assuming population sizes as given in either the best-case scenario (upper lines) or the worst-case scenario
(lower lines) in Fig. 7.1, and either random mortality of juveniles prior to reproduction
(solid lines) or loss of entire litters (dashed lines).

Figure 7.4 shows the expected loss of heterozygosity from the Meeteetse population resulting from genetic drift since 1935. Estimates were
made under the assumption of random mortality (solid lines) and
whole-litter mortality (dashed lines), and for the best-case scenario (upper lines) and worst-case scenario (lower lines) by iteratively applying
Equation 7.2, with an exponent of $t = \frac{2}{3}$, to the annual effective population sizes given in Figure 7.3. The exponent corrects for the assumption
of a 1.5-year generation time. The 1935 heterozygosity was set at 1.0,
and the heterozygosity values along the y axis are therefore proportions
remaining of the 1935 heterozygosity. It is reasonable to guess that the
primal Meeteetse population had about 1% heterozygosity (see also
O'Brien et al., chap. 3, this volume). This value is lower than is observed
for most mammals (Nevo 1978) but is not atypical of carnivores (Kilpatrick et al. 1986) and may be greater than is typical of other mustelids
(Simonsen 1982). Due to the low population density and low dispersal of
ferrets, 1935 heterozygosity may well have been less than 1%.

The heterozygosities in Figure 7.4 were augmented each generation

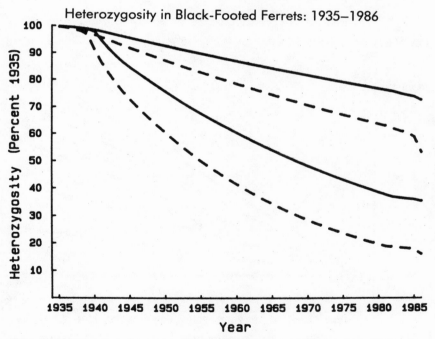

Figure 7.4. Heterozygosity expected to be maintained in the Meeteetse population of
ferrets (BFFs) under the conditions of Fig. 7.3, relative to an initial (assumed) heterozygosity of 1.0 (or 100% on the y axis) in 1935.

to account for new variation expected to be introduced by mutation. The mutation rate was assumed to be 10^{-6} per gene per generation (thus 0.000002 mutations would occur at a genetic locus per diploid individual per generation). A mutation rate of 10^{-6} may be generous, considering that any markedly deleterious mutations would be eliminated by natural selection and would not contribute to heterozygosity. Incrementing heterozygosity by just 0.0002% per generation, mutation had a negligible effect on the estimated heterozygosities in the 1935–86 interval.

Projected Heterozygosity under Various Population Growth Rates, 1986–2036

Population sizes for ferrets were projected through the next 50 years under assumptions of 0%, 5%, 10%, 15%, 20%, and 25% annual growth rates (Fig. 7.5). Such population growth rates should be attainable in captivity, and even wild populations could grow at rapid rates if there

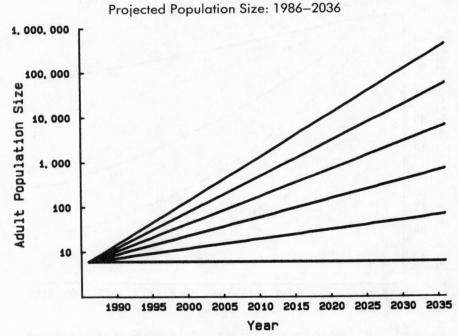

Figure 7.5. Projected population sizes of ferrets, starting with 6 ferrets in 1986, with varying rates of population growth per year.

were extensive prairie dog complexes available into which the ferret populations could expand. The maximum 1-year increase observed in the Meeteetse population between 1982 and 1986 was 26%.

The expected fate of heterozygosity was calculated for each projected growth rate, with appropriate adjustments being made for a 1 : 2 sex ratio, variance in population size (CV set at 0.3003, as above), and variance in family size with either random juvenile survival (Fig. 7.6) or family survival (Fig. 7.7). The percentage of heterozygosity was incremented by 0.0002 each generation, under the assumption that the mutation rate would be 10^{-6}.

Early projected generations continue to lose heterozygosity because of genetic drift, while heterozygosity begins to increase, almost imperceptibly, in later generations as mutation restores genetic variation. After the population reaches sufficient size so that genetic drift is negligible (on the order of thousands), it would take several thousand genera-

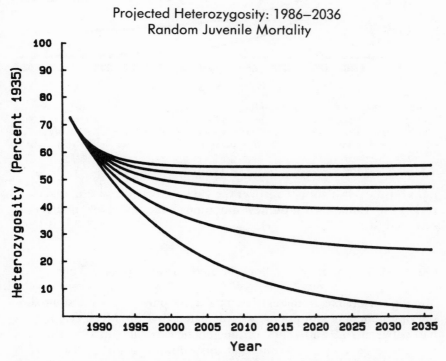

Figure 7.6. Projected levels of heterozygosity in ferrets from 1986 to 2036, assuming random prereproductive mortality and an initial level of heterozygosity of 0.73% (the best-case outcome in Fig. 7.4), with the varying rates of population growth in Fig. 7.5. (Percentage of heterozygosity is relative to an assumed level of 1.0 [or 100% on the y axis] in 1935.)

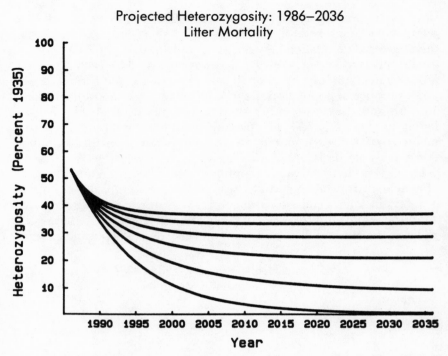

Projected Heterozygosity: 1986–2036
Litter Mortality

Figure 7.7 Projected levels of heterozygosity in ferrets from 1986 to 2036, assuming prereproductive mortality of entire litters and an initial level of heterozygosity of 0.53% (the best-case outcome in Fig. 7.4), with the varying rates of population growth in Fig. 7.5. (Percentage of heterozygosity is relative to an assumed level of 1.0 [or 100% on the y axis] in 1935.)

tions for the population to recover the 1% heterozygosity estimated to be present in 1935, even if all new mutations were selectively neutral (as assumed in Figs. 7.6 and 7.7).

Genetic Considerations in Black-Footed Ferret Recovery Plans

Making reasonable estimates of demographic parameters and some simplifying genetic assumptions, we calculated that the Meeteetse ferrets had between 15.9 and 72.6% of the genetic variability in 1986 that was present in 1935 (Fig. 7.4) and that this would be about 0.16 to 0.73% heterozygosity. Our best estimate of the heterozygosity remaining in the Meeteetse ferrets might be the average of the outcomes shown in Figure 7.4, 44.2% of the 1935 level. Alternatively, we might assume that from

1940 to 1981 the population remained at a mean of 31.75 ferrets, the average size from 1982 to 1985. Given this average case, the Meeteetse population would have declined to 61% of the 1935 heterozygosity if there were random juvenile mortality and 39.7% of 1935 heterozygosity under whole-litter mortality.

Our upper-bound estimate of the remaining heterozygosity (73% of initial heterozygosity in 33 generations in the best-case scenario with random mortality of offspring) resulted from approximately a 1% loss per generation, the maximal loss considered acceptable for short-term breeding programs with domestic livestock (Franklin 1980). The rate of decline in the worst-case scenario with family mortality, about 5% loss per generation, is about the maximal loss considered unlikely to cause genetic problems in one generation in human and domestic populations. (Offspring produced by the mating of first cousins have an inbreeding coefficient of .0625, that is, they are expected to have 6.25% less heterozygosity than noninbred individuals in the population.) In either scenario, low levels of heterozygosity remain, however. Thus, further losses of genetic variability could still be deleterious to viability and fecundity (inbreeding depression), even though the prolonged nature of the population bottleneck should have provided opportunity for natural selection to eliminate many deleterious recessive alleles.

Because of low genetic variation, response to selection (artificial or natural) in ferrets would be much less efficient today than in 1935. The effectiveness with which a shortage of prairie dogs and then an outbreak of distemper reduced the ferret population in 1985 perhaps indicates an inability of the ferret population to adapt rapidly to environmental stresses. Although the loss of heterozygosity has not been extreme, initially rare genetic variants would be very unlikely to have persisted through the bottleneck. Thus, alleles that are highly adaptive only when the population is subjected to unusual stresses would have been lost from the ferret population and unavailable for adaptive response when those conditions recur.

If the Meeteetse population were to remain at a size typical of that maintained from 1940 to 1985 (N_e perhaps 25), it would reach mutation-drift balance when heterozygosity had declined to about 0.01%. Only an average of 1 locus (out of an estimated 10,000 structural loci) in the genome would be expected to be heterozygous in each individual, and those heterozygous loci would be mostly due to recent mutations not yet lost by drift, not due to persistent polymorphisms in the population. Although the population probably still contains low levels of variability (Fig. 7.4), because of the low current population numbers further losses will be unavoidable even if the population grows rapidly (Figs. 7.6 and

7.7). If the history of Meeteetse ferrets is closer to our optimistic scenario and population growth is rapid, then perhaps as much as half of the pre-1935 heterozygosity can be preserved. If the pessimistic scenarios more closely approximate the history of the Meeteetse ferrets, then the ferrets will retain only very minimal genetic variation after the current bottleneck ends, even if recovery is rapid after 1986.

It is unlikely that a long-term viable population of ferrets currently exists in the wild at Meeteetse or elsewhere. The captive population, with at most 9 unrelated individuals and their offspring, will probably have to form the primary founding stock for all future captive and wild ferret populations. Ideally, 10 to 20 wild-caught pairs should be used to begin a captive propagation effort to assure that the captive stock is a good genetic representation of the wild population (containing 97.5 to 98.75% of the heterozygosity present in the source population). It is certainly too late to accomplish that ideal for the Meeteetse ferrets, as it probably has been since the 1930s. If any ferrets remain in the wild around Meeteetse, it would be beneficial to bring them into captivity. The addition of just 1 more male to the captive population in 1987 increased the effective size of the population of possibly unrelated adults from 7.5 (3 males, 5 females) to 8.9 (4 males, 5 females). Another unrelated male could bring N_e to 10.

Although searches for ferrets in Montana, Colorado, and elsewhere have been unsuccessful thus far, we can still hope that another remnant population will be found. If a second population is discovered, distant enough to have assured isolation from the Meeteetse population since the 1930s or earlier, the 2 populations should probably not be fully merged. Both populations would likely have become adapted to very low heterozygosity (unless the newly discovered population were large), and mixing of the 2 might produce gene combinations poorly adapted to either, or even any, locality. After both populations had recovered to numbers sufficient to be out of immediate danger of extinction, a subset of each could be combined to produce a third population. The mixed population would be greater in variability and could be used for attempted introductions to other suitable habitats.

Immediate management of the captive stock can probably safely ignore genetic constraints and may have no option to do otherwise. Unless more ferrets are discovered and brought into captivity, soon all captive-born ferrets will be related to all others, and inbreeding will be unavoidable. Due to the probable lack of substantial genetic variation in ferrets and to the opportunity for selection to have removed recessive deleterious alleles from the Meeteetse population over the past 50 years, further moderate inbreeding would be unlikely to cause new and serious genetic problems. Behavioral avoidance of close inbreeding would be more likely to hinder captive propagation efforts. Clearly, demographic management of the population is far more important than genetic management at this time. Moreover, achieving a high rate of increase in the population is the most effective means for slowing further losses in genetic variability (Figs. 7.6 and 7.7).

The lack of variation and very small population size should also alleviate any concerns about genetic domestication of the ferrets through inadvertent selection of the captive population. Little variation exists on which selection could act, and random drift is a much stronger evolutionary force on very small populations than is selection (Lacy 1987). Certainly, no effort should be made to remove ferrets that appear to have genetic defects from the breeding population. Such artificial selection would accelerate the loss of genetic variability (at many loci, not just those involved in the genetic defect) and slow the rate of population growth. Genetic problems, which may be revealed by inbreeding, can be dealt with (if deemed desireable) after the population is out of immediate danger of extinction. Behavioral changes that occur in captivity would be more likely than genetic changes to cause problems for captive propagation and eventual reintroduction programs (Richardson et al. 1986).

Because genetic variability is probably low, management plans should be cognizant of the possibly extreme sensitivity of ferrets to diseases, parasites, or any environmental stresses. Multiple captive and wild stocks should be created as soon as sufficient numbers of animals exist to form demographically viable populations in order to minimize the probability of a catastrophic loss of all ferrets due to disease, parasites, or inadvertently inappropriate management. Any deleterious effects of increased genetic drift occurring in, for example, 3 populations of 10 animals, rather than 1 population of 30, would be minimal. If moderate genetic variability does still exist in the Meeteetse ferrets, multiple stocks would also have the advantage of minimizing problems associated with domestication or adaptation to the captive environment resulting from unintentional selection. Selection would probably favor different traits in different captive stocks and drift would fix different traits by chance, so that more variation would be maintained overall among multiple stocks than within a single stock (Lacy 1987).

Once a captive population exceeds about 20 ferrets, the short-term genetic advantage of maintaining more animals is small. For example, a population of 20 would lose about 2.5% of its heterozygosity (if it had any to start with) per generation, while a population of 25 would lose about 2.0% of its heterozygosity per generation. To assure demographic stability, a much larger population may be necessary. If viable wild populations are found, captive animals should probably not be released into those populations unless releases are needed to maintain the demographic stability of the wild populations. The genetic effects of introducing a few animals to a stable wild population would be minimally beneficial and could be harmful.

Additional wild and captive populations could be started with as few as 2 pairs (the full siblings produced by a single pair would probably avoid intermating), if all translocated ferrets survived to reproduce. Three or 4 pairs would provide more assurance that enough translocated animals would survive to establish the new population. Animals could be added every year or 2 until the population reached a demographically safe size. Groves and Clark (1986) estimated that 15 female ferrets would be needed to achieve an 80% probability that a re-established population would not go extinct because of demographic stochasticity, and that 200 ferrets might constitute a minimum viable population that would maintain sufficient genetic variability to make extinction unlikely. After several populations are established at demographically and genetically safe levels (perhaps 200 animals), occasional migration between populations could reduce inbreeding, but should be conducted with caution. The movement of just 1 animal per generation would lessen inbreeding (with uncertain benefits in the case of ferrets), though at the risk of losing the migrant.

Long-term plans should aim for the establishment of 10 or more wild populations of ferrets, supplemented and ensured by an active captive propagation program at multiple sites. The probability that a new, neutral mutation will eventually drift to fixation within a population is equal to its initial frequency, $1/(2N)$. Therefore, new mutations are more likely to become established in at least 1 local population if the species is kept fragmented into a number of small (but not so small as to be demographically unstable) demes. Occasional migration between populations, on the order of 1 animal moved per generation, will allow random and adaptive divergence between local populations but still provide opportunity for the spread of new genetic variants.

References

Anderson, E., et al. 1986. Paleobiology, biogeography, and systematics of the black-footed ferret, *Mustela nigripes* (Audubon and Bachman, 1851). *Great Basin Nat. Memoirs* 8:11–62.

Bonnell, M. L., and R. K. Selander. 1974. Elephant seals: Genetic variation and near extinction. *Science* 184:908–9.

Clark, T. W., et al. 1986. Description and history of the Meeteetse black-footed ferret environment. *Great Basin Nat. Memoirs* 8:72–84.

Crow, J. F., and M. Kimura. 1970. *An introduction to population genetics theory.* New York: Harper and Row.

Crow, J. F., and N. E. Morton. 1955. Measurement of gene frequency drift in small populations. *Evolution* 9:202–14.

Flath, D. L., and T. W. Clark. 1986. Historic status of black-footed ferret habitat in Montana. *Great Basin Nat. Memoirs* 8:63–71.

Forrest, S. C., et al. 1985a. Black-footed ferret habitat: Some management and reintroduction considerations. *Wyoming BLM Wildlife Technical Bulletin* no. 2. Cheyenne, Wyo.: U.S. Bureau of Land Management.

Forrest, S. C., et al. 1985b. Life history characteristics of the genus *Mustela,* with special reference to the black-footed ferret, *Mustela nigripes. Proc. black-footed ferret workshop,* ed. S. H. Anderson and D. B. Inkley, 1–23. Cheyenne, Wyo.: Wyoming Game and Fish Dept.

Forrest, S. C., et al. 1988. Population attributes for the black-footed ferret (*Mustela nigripes*) at Meeteetse, Wyoming, 1981–1985. *J. Mamm.* 69:261–73.

Franklin, I. R. 1980. Evolutionary change in small populations. In *Conservation biology: An evolutionary-ecological perspective,* ed. M. E. Soule and B. A. Wilcox, 135–50. Sunderland, Mass.: Sinauer Assoc.

Groves, C. R., and T. W. Clark. 1986. Determining minimum population size for recovery of the black-footed ferret. *Great Basin Nat. Memoirs* 8:150–59.

Hoppe, K. M., et al. 1984. Biochemical variability in a population of beaver. *J. Mamm.* 65:673–75.

Kilpatrick, C. W., et al. 1986. Estimating genetic variation in the black-footed ferret—a first attempt. *Great Basin Nat. Memoirs* 8:145–49.

Lacy, R. C. 1987. Loss of genetic diversity from managed populations: Interacting effects of drift, mutation, immigration, selection, and population subdivision. *Conservation Biology* 1:143–58.

Lidicker, W. Z., and F. C. McCollum. 1981. Genetic variation in the sea otter, (*Enhydra lutris*). *Bien. Conf. Biol. Mar. Mammals,* 4th, San Francisco, California, 14–18 December 1981.

Nei, M., et al. 1975. The bottleneck effect and genetic variability in populations. *Evolution* 29:1–10.

Nevo, E. 1978. Genetic variation in natural populations: Patterns and theory. *Theoret. Popul. Biol.* 13:121–77.

O'Brien, S. J., et al. 1983. The cheetah is depauperate in genetic variation. *Science* 221:459–62.

O'Brien, S. J., et al. 1985. Genetic basis for species vulnerability in the cheetah. *Science* 227:1428–34.

Ralls, K., et al. 1983. Genetic diversity in California sea otters: Theoretical considerations and management implications. *Biol. Conserv.* 25:209–32.

Richardson, L., et al. 1986. Black-footed ferret recovery: A discussion of some options and considerations. *Great Basin Nat. Memoirs* 8:161–84.

Simonsen, V. 1982. Electrophoretic variation in large mammals. II. The red fox (*Vulpes vulpes*), the stoat (*Mustela erminea*), the weasel (*Mustela nivalis*), the polecat (*Mustela putorius*), the pine marten (*Martes martes*), the beech marten (*Martes foina*), and the badger (*Meles meles*). *Hereditas* 96:299–305.

PART THREE

Reproduction

8

BRUCE D. MURPHY

Reproductive Physiology of Female Mustelids

In the order Carnivora, the family Mustelidae consists of a moderately diverse group of animals, varying in size from the large sea otter (*Enhydra lutris*), weighing 25 kg or more, to the smallest carnivore, the least weasel, (*Mustela nivalis*), which rarely exceeds 250 grams. At least 3 species, the European ferret (*Mustela putorius*), the mink (*Mustela vison*), and the sable (*Martes zibellina*), have been raised extensively in captivity. The ferret alone is gentle enough to qualify as a domestic animal and has been the subject of domestication in Europe for at least 2,000 years (Hammond and Chesterman 1972). The first ferrets were imported to the United States in 1875 (Ryland and Gorham 1978). It is not surprising that most information on the reproductive physiology of mustelids has been assembled from studies of the ferret and the mink. The reproductive biology of 2 wild species is well known: the western spotted skunk (*Spilogale putoris*) because of the extensive investigations of R. A. Mead; and the European badger (*Meles meles*) because of the work of R. Canivenc. There is a dearth of descriptive or experimental investigations of any other mustelid. For these reasons this chapter will emphasize the mink and ferret, drawing examples from other species when appropriate.

The events of reproduction in any mammal can be summarized to include: puberty and the associated initiation of development of ovarian follicles to the preovulatory state, estrus, ovulation, embryo migration to the uterus, embryo implantation and development, ascendancy and demise of the corpus luteum, and parturition. This chapter will address each of the topics, with the exception of parturition, about which little information is currently available in carnivores. Virtually all mustelids breed seasonally, and puberty usually coincides with reattainment of breeding capability in adult animals. The mechanisms by which pho-

The author thanks S. Johannesen for her help in preparing this manuscript.

toperiod influences puberty and seasonal breeding are discussed by
Herbert (chap. 10, this volume).

Intervention into the natural pattern of mustelid reproduction has
been carried out to a limited degree. Manipulation of photoperiod can
induce mink to breed out of season (Travis et al. 1978). Hormonal
treatment has been used to bring about follicular development in both
mink and ferrets, and alteration of luteal function can be achieved by
treatment with neuroleptic drugs. This presentation recounts potential
and actual methods of the modification of reproductive function and
their physiological bases in female mustelids.

Reproductive Physiology of Mustelids

The process of follicular development in mammals has been reviewed
by Murphy (1986). It begins before birth with the mitotic division of the
gametic cells. All of the ova that a particular female mammal will have
are present before birth. In most species, primordial follicles develop
during the prenatal era, a process consisting of aggregation of the initial
follicular (granulosa) cells about the oocyte, which is arrested in the state
of meiotic prophase. The primordial follicle is avascular and is separated from the remainder of the ovary by means of a basal lamina. The
adjacent ovarian interstitial cells differentiate into the theca interna, a
highly vascularized and endocrinologically important layer of cells.

Primordial follicles comprise the pool of undeveloped ova in the ovary. The next stage of development, to primary follicles, is characterized by proliferation of the granulosa cells. Primordial follicles have
been shown to develop to primary follicles during all stages of the reproductive life span of mammals, including prenatal, prepubertal, and
anestrous periods and pregnancy. This development is most pronounced in proestrous females. Primary ovarian follicles appear
throughout the year in anestrous and prepubertal mink, with a conspic-
uous increase in numbers occurring during the breeding season (Enders
1952). Initiation of follicular development, suggested by the presence of
primary follicles, occurs in hypophysectomized mink (Murphy et al.
1980) and ferrets, suggesting that the pituitary is not required for entry
into the pool of developing follicles (Murphy 1979a). The endocrine
bases for reinitiation of follicular development are not known. Once a
follicle enters the growing pool by undergoing the transformation from
primordial to primary follicle, only 2 paths are open: development to a
preovulatory follicle or degeneration (atresia).

Follicular progression beyond the primary stage consists of proliferation of granulosa cells and the formation of a fluid-filled cavity called the antrum. This series of events is associated with proestrus of the breeding season. During this period, antral follicles increase in size to form preovulatory follicles in mink (Hansson 1947) and ferrets (Robinson 1918). While there are isolated examples of development of antral follicles in anestrous mink (Enders 1952), it was not observed to occur following hypophysectomy in either mink (Murphy et al. 1980) or ferrets (Murphy 1979a). This suggests that pituitary gonadotropins are required for follicular maturation, as in other species.

In both ferrets and mink, ovulation is induced by copulation (discussed below) and can occur at any time after the animals have attained estrus (Venge 1959; Robinson 1918). Mink have a 4- to 5-week reproductive season during which ovulation is induced and successive crops of ovarian follicles develop at 6- or 7-day intervals (Murphy 1982). In ferrets, estrus can persist for up to 5 months, but once ovulation is induced, pregnancy or pseudopregnancy ensues (Hammond and Marshall 1930). The follicular events that transpire during an unrequited estrus in these 2 species are unknown. There are 2 possibilities: (1) the initial group of follicles persists, or (2) follicular development and atresia are continuous and overlapping so that there is a recent crop of follicles available for ovulation whenever copulation might occur. Histological observations suggest the latter hypothesis is more tenable. Waves of follicles develop in the absence of mating in mink (Enders 1952) and ferrets (Hartman 1939). The interval required for development from primordial to preovulatory follicle has not been reported for any mustelid. The interval required for development of a new crop of follicles following ovulation is 6 days in mink (Hansson 1947). It is postulated that the time required for a wave of follicles to develop prior to induced ovulation in ferrets and mink is approximately 6 days. The waves of development are to represent the succession of follicles from the preantral or early antral stages to the preovulatory stage. This is presumed to occur under the influence of gonadotropins as demonstrated in the rat (Hirshfield and Midgely 1978). Histological evidence suggests that the number of developing follicles is greater in the ovaries of proestrous mink than in those of anestrous mink (Enders 1952).

Moon et al. (1976) have observed that antral and preovulatory follicles secrete estrogens, which are produced by granulosa cells under the influence of follicle stimulating hormone (FSH). Domestic ferrets are unusual in that preovulatory follicles and attendant estrogen secretion are present for the long duration of estrus in unbred animals. A consequence of prolonged estrus is an estrogen-induced aplastic anemia characterized by leukopenia, thrombocytopenia, and hypocellular bone marrow, which is usually fatal (Kociba and Caputo 1981; Ryland 1982). It is therefore advisable to terminate estrus in domestic ferrets by means of mating or induction of ovulation.

Estrus occurs in March in ferrets maintained under natural photoperiod in the Northern Hemisphere (Ryland and Gorham 1978). Attainment of breeding condition appears to be independent of body weight, although there may be a minimum weight (420 gm) below which estrus does not occur (Donovan 1986). As with a number of other species of mammals, captive ferrets display an increased frequency of activity with the onset of estrus (Donovan 1986).

During estrus there is vulval edema in both ferrets and mink. In ferrets, where the effect is most pronounced, maximal size of the vulva is attained approximately 1 month after the first changes are noticeable in anestrous animals (Hammond and Chesterman 1972). The same authors report an interval of only 10 days from initiation of edema until maximal swelling is achieved in animals undergoing second estrus in a year. Estrogens from the developing follicles are responsible for vulval swelling, an effect antagonized by progesterone (Marshall and Hammond 1945). In general, animals with vulval diameters greater than 1 cm have been found to successfully mate (Murphy, unpubl.). Vulval edema, while much less pronounced, has been used to indicate readiness to mate in mink (Travis et al. 1978).

The feedback control of luteinizing hormone (LH) secretion has been studied in estrous ferrets by Tritt et al. (1986), who demonstrated that LH secretion during estrus does not exhibit the usual mammalian pattern of pulsatile release. They confirmed the observation of Donovan and ter Haar (1977) and Carroll et al. (1985) that modest increases in circulating LH result from administration of gonadotropin releasing hormone (GnRH). Tritt et al. (1986) further demonstrated that the tamoxifen elevated circulating LH, suggesting that estrogen is involved in feedback control of LH during estrus.

In some mustelids, ovulation is induced by copulation. In others, including the spotted skunk, it occurs spontaneously (Greensides and Mead 1973). Robinson (1918) was the first to demonstrate induced ovulation in the ferret. Early studies determined the time required from the initiation of coital stimulation until expulsion of the ovum in both mink and ferret. In mink, ovulations begin to occur at approximately 36 hours post-coitum (Enders 1952; Hansson 1947). Ferrets have been shown to ovulate beginning approximately 30 hours after initial intromission (Hammond and Walton 1934).

During mating, which can be a protracted event in both mink and ferrets, the male grips the dorsal surface of the female's neck and makes repeated pelvic thrusts until intromission is achieved (Carroll et al. 1985). Baum and Schretlen (1975) reported that the duration of single intromissions in ferrets varied from 2.5 to 151.5 minutes. Similar intervals have been reported for mink (Garcia-Mata 1982). Intromission was shown to be requisite to induction of ovulation in ferrets, since neck gripping and pelvic thrusting produced no effect (Carroll et al. 1985). The same authors determined that intromission persisting for only 1 minute induced ovulation.

Hormonal measurements have been made during induced ovulation in rabbits (Dufy-Barbe et al. 1973) and cats (Concannon et al. 1980). In both species, there is a nearly instant increase in circulating LH concentrations following intromission. LH values are elevated 20- to 40-fold within 10 minutes, and peak values are attained by 0.5 to 2 hours. Studies of ferrets by Hill and Parkes (1932) demonstrated that hypophysectomy within 1 hour and 50 minutes after mating did not prevent ovulation, suggesting a similar pattern of LH release. Nevertheless, only modest increases in LH were observed following intromission and only a 3- to 4-fold elevation ensued (Fig. 8.1), peaking 3 to 10 hours later (Carroll et al. 1985). These observations suggest that only a brief interval of exposure to elevated LH may be required to induce ovulation. Events such as neck gripping and pelvic thrusting in the absence of mating did not produce increases in LH (Carroll et al. 1985). No hormonal data appear to be available on mink ovulation induced by coitus. However, 7 minutes of cervical stimulation with a 4-mm glass rod resulted in LH release in 3 of 5 female mink (Murphy and Shepstone, unpubl.). In the animals that responded, LH release began within 20 minutes of initiation of cervical stimulation, and LH was above basal concentrations at 6 hours (Fig. 8.2).

The sequelae to LH secretion and the manner in which LH induces expulsion of ova remain enigmatic in mustelids. The number of ova released by mink and ferret has not been extensively studied. Carroll et al. (1985) recorded between 5 and 11 corpora lutea resulting from a single mating in ferrets. Hansson (1947) reported a range of 4 to 15 ova were released following individual matings in mink.

It has not been conclusively shown where fertilization occurs in either the mink or the ferret. The discovery of fertilized ova in serially sectioned oviducts of mink 53 hours after mating led Enders (1952) to conclude that fertilization takes place within the fimbriated or more distal portions of the oviduct. Hamilton (1934) described a series of ferret ova in early stages of fertilization which were found in the oviduct 40 to 41 hours after insemination. Robinson (1918) concluded that fertilization in the ferret takes place in the middle third of the oviduct.

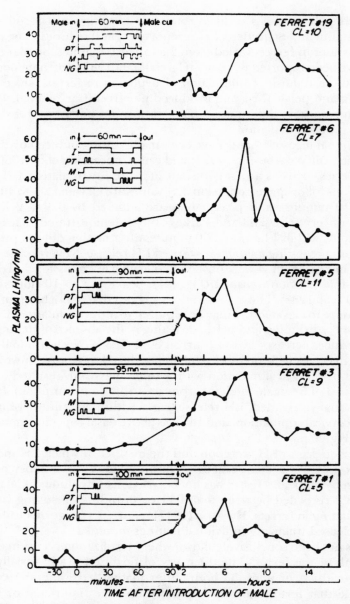

Figure 8.1. Patterns of plasma LH levels in female ferrets in estrus. Samples were taken prior to and during the introduction of a male ferret in breeding condition. The inserts record the occurrence and duration of male copulatory behavior including NG (neck grip), M (mount), PT (pelvic thrusting), I (intromission). (Reprinted with permission from Carroll et al. 1985. *Biol. Reprod.* 32:925.)

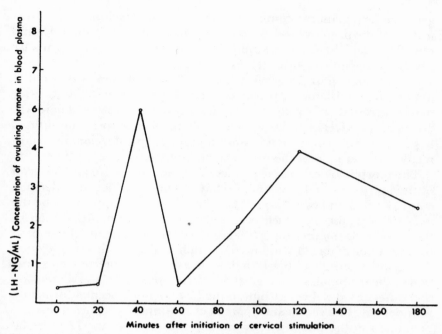

Figure 8.2. LH in plasma in a restrained ranch mink subjected to 7 minutes of cervical stimulation with a 4-mm glass rod (Murphy and Shepstone, unpubl. observ.).

Ryland and Gorham (1978) write that sperm may penetrate the ferret ovum before it is released from the follicle, although no experimental evidence is cited in support of this assertion.

Embryonic development in the ferret following fertilization has been described in detail by Hamilton (1934), who reported that the initial cleavage stages can be found 51 hours after insemination. There was considerable variation in the time at which the 2-cell stage appears. The 4-cell stage was collected from a number of animals between 64 and 72 hours after copulation. Subsequent divisions to 16-cell morulae were observed in the oviduct before 120 hours in Hamilton's (1934) study and were confirmed by Daniel (1970). Blastocysts were present in the ferret oviduct within 5 days after fertilization (Hamilton 1934).

Mink embryology is less well documented. In mink, embryos progress rapidly down the oviduct during the first 12 hours following ovulation and fertilization, and much more slowly thereafter (Enders 1952). Blastocysts arrive in the upper reaches of the uterine horns by the 6th or 7th day after mating (Hansson 1947; Enders 1952).

Upon arrival in the uterus, the fate of embryos of the 2 species diverges. The ferret embryo continues to expand and develop, and the formation of the primitive streak may ensue prior to implantation (Hamilton 1934; Daniel 1970). The mink blastocyst enters a state of diapause which can persist for weeks (Hansson 1947).

In a recent report, Rider and Heap (1986) presented evidence that immunoneutralization of progesterone in ferrets from 72 hours after mating arrested embryonic development at the 4-cell stage. There was no apparent effect of antiprogesterone on tubal transport. The authors hypothesized that passive immunization interrupts development indirectly, via alteration of the oviductal environment.

The morphological characteristics of the attachment to and invasion of the uterus by the ferret embryo have been the subject of a detailed study (Enders and Schlafke 1972). The initial event appears to be the swelling of embryos, which occurs day 9 postcoitum. Apposition of the embryo to the uterine epithelium and its consequent adhesion on the afternoon of day 11 is followed by trophoblastic evagination through the acellular capsule of the 12th day. Invasion of the uterus by insinuation of the trophoblast between endometrial epithelial cells follows soon thereafter. By day 14 postcoitum, the first step in placental formation is underway. In a recent study, Mead et al. (unpubl.) reported that implantation in the ferret uterus was accompanied on day 12 of gestation by highly localized increases in vascular permeability.

The process of implantation has been less well described in the mink because, due to embryonic diapause, the timing of attachment is highly variable. There is, however, good morphological information on the embryo of the spotted skunk in diapause (Enders et al. 1986) and on embryo implantation with the same species (Sinha and Mead 1976). These reports suggest that the process is qualitatively similar to implantation in the ferret.

Embryo implantation does not occur in ferrets if the ovaries are removed between days 6 and 12 of gestation (Wu and Chang 1972, 1973). Embryo attachment in that species depends on the presence of an intact pituitary (Murphy 1979a). Endocrine requirements for implantation of ferret embryos include progesterone (Rider and Heap 1986) and some other luteal factor secreted between days 6 and 8 of gestation (Foresman and Mead 1978). Despite pronounced activity of the aromatase enzyme which induces estrogen synthesis (Mead and Swannack 1980), the evidence gathered to date does not support the concept that the unknown luteal factor is an estrogen (Murphy and Mead 1976; Mead and McCrae, 1982). The pituitary hormone prolactin (PRL), administered daily, will bring about embryo implantation in hypophysectomized ferrets, presumably by inducing secretion of the appropriate hormones from the corpora lutea (Murphy 1979a). Uterine prostaglandins have been implicated in local control of implantation in rats (Kennedy 1977). Inhibition of prostaglandin synthesis by indomethacin had no effect on the increases in vascular permeability which precede implantation in the ferret (Mead et al. 1987).

In mink, embryonic diapause of variable length is terminated by implantation (Hansson 1947; Enders 1952). Hormonal correlates parallel those in the ferret. Progesterone and other factors from the corpus luteum are required for implantation (Murphy et al. 1983), and PRL, presumably acting as a luteotropin, will induce implantation in intact (Papke 1980) and hypophysectomized mink (Murphy et al. 1981).

Ovulation and collapse of the follicle initiates development of the corpus luteum, the organ responsible for maintenance of pregnancy in both mink and ferrets. The corpus luteum is believed to be required throughout gestation, although experimental evidence in ferrets is confined to the first trimester. Ovariectomy will result in failure to implant in the ferret and loss of embryos after implantation (Wu and Chang 1972). In mink, ovariectomy precludes implantation and results in resorption of implanted embryos (Murphy et al. 1983).

The course of progesterone during gestation in mink and ferrets (Fig. 8.3) is similar in that levels are initially low, begin to increase prior to implantation, and peak during early postimplantation gestation (Mur-

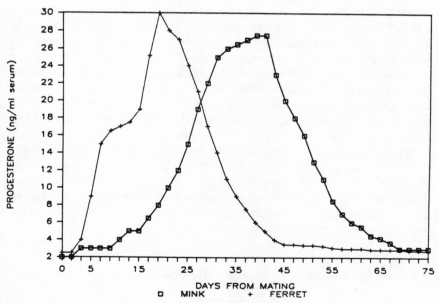

Figure 8.3. Representative profiles of progesterone concentrations in mink and ferret following ovulation induced by copulation.

phy and Moger 1977; Heap and Hammond 1974). The length of postimplantation gestation in both species is approximately 30 days, resulting in a total gestation length in the ferret of 42 days. Because of the phenomenon of delayed implantation, the length of mink gestation can vary greatly. The lower limit appears to be about 40 days, and individual pregnancies have exceeded 75 days under some experimental conditions (Murphy and James 1974).

In contrast to ungulates and rodents, in which regression of the corpus luteum is abrupt, mustelid and other carnivores display a protracted decline in circulating progesterone. Luteal cells retain an active appearance, and progesterone levels do not reach basal levels until 1 or more weeks after parturition (Moller 1973). The conceptus has no apparent effect on the duration or magnitude of the luteal phase in the mink and ferret because pregnancy and pseudopregnancy are indistinguishable (Hammond and Marshall 1930; Heap and Hammond 1974; Agu et al. 1986). By the same token, no control can be ascribed to the uterus in the ferret, as its excision has no effect on the life span of the corpus luteum (Deanesly 1967). The proximate factor which induces luteal regression in either species is not known.

Luteal morphology and progesterone secretion require an intact pituitary (Figs. 8.4a and 8.4b) in both mink and ferrets (Hill and Parkes 1932; McPhail 1935; Donovan 1967; Murphy et al. 1980). Early studies suggested that PRL was the sole luteotropic agent in the ferret (Donovan 1967; Murphy 1979a), as hypophysectomy, followed by PRL replacement, maintained luteal function. More recently, Agu et al. (1986) presented 3 lines of evidence indicating that the ferret corpus luteum was governed by a luteotropic complex rather than by PRL alone. The study showed that luteal function could be disrupted during the 2 weeks following ovulation by treatment with the dopamine agonist bromocriptine, confirming the requirement for PRL. Antiserum against LH or GnRH, treatments which abolish circulating LH, reduced circulating progesterone. Daily treatment with human chorionic gonadotropin (hCG), which interferes with luteal function by reducing the numbers of the LH receptors on luteal cells, also disrupted progesterone secretion. Taken together, these observations suggest that during the first third of gestation the ferret CL, while dependent primarily on PRL, has a requirement for LH as well.

Potential for Alteration of Reproductive Function in Mustelids

As noted above, development of follicles to the preovulatory state in mammals occurs as the result of stimulation of the preantral or early

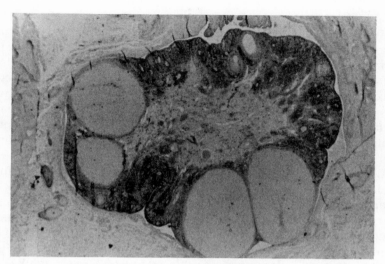

Figure 8.4a. Photomicrograph of a histological section of the ovary of an intact ferret on the 15th day after mating. Note the prominent corpora lutea and follicles.

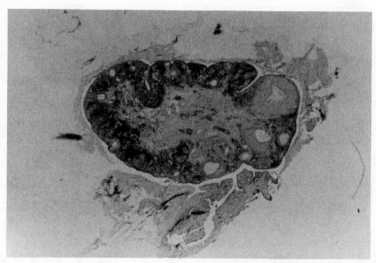

Figure 8.4b. Section of an ovary from a hypophysectomized ferret 15 days after mating and 10 days after removal of the pituitary. Note luteal atrophy.

antral follicles by the gonadotropic hormones, FSH and LH (Murphy, 1986). It is possible to induce superovulation in cattle and sheep and precocious puberty in pigs. The hormones commonly employed are (1) commercial preparations of FSH, which usually contain more LH than FSH (Lindsell et al. 1986), (2) menopausal gonadotropins, which have approximately equal amounts of FSH and LH (Murphy et al. 1984), and (3) equine chorionic gonadotropin (eCG), also known as pregnant mare serum gonadotropin or PMSG (Murphy et al. 1987). The latter compound has both LH and FSH activity in mammals other than the horse (Gonzalez-Mencio et al. 1978).

Only 1 of these compounds, eCG, has been used to induce follicular development in mustelids. Hammond (1952) injected pelter mink in December with 100 IU of eCG followed by hCG some 9 days later and determined that follicular development and ovulation ensued. A similar quantity of eCG was injected into 226 mink that had failed to breed by late in the breeding season (March 20) on an Ontario mink ranch (Murphy et al., unpubl.). Breeding was achieved in 84.5% of the mink (Fig. 8.5), and 49.7% of the bred females whelped, averaging 3.7 kits per litter. The untreated animals that had bred normally had a whelping success of 84.4% and mean litter size of 4.19 kits per female whelping. The reasons that the mink failed to mate are not known. Laparotomies were not performed, and it is not known if eCG induced follicular

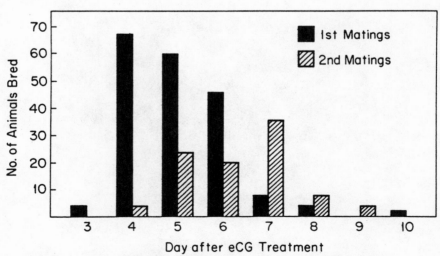

Figure 8.5. Occurrence of mating in female mink that had previously been unreceptive and were treated with 100 IU equine chorionic gonadotropin (eCG).

development. The evidence of successful mating suggests that this was the case.

Domestic ferrets in anestrus have been treated with eCG (100 IU) followed 72 hours later by 100 IU of hCG to induce follicular development and pseudopregnancy (Agu et al. 1986). No differences were detectable in either the number of ovulations in anestrous ferrets following eCG-hCG treatment or the induction of ovulation in estrous ferrets treated with hCG. There were no apparent differences in luteal function *in vivo* or *in vitro*. Vulval swelling did not consistently appear in anestrous ferrets following eCG treatment, in spite of the development of preovulatory follicles and ovulation. This may have been due to the brief interval (72 hours) between hormone treatment and ovulation.

Ovulation in most mustelids is induced by copulation, which appears to result in release of LH. The success of artificial insemination depends on concomitant induction of ovulation, and cervical stimulation may be an impractical means of achieving this purpose. The human placental hormone with LH activity, hCG, has been used extensively to induce ovulation in mink at doses of 40 and 50 IU per animal (Adams 1976). In the domestic cat, also an induced ovulator, the response to hCG is dose dependent, and the maximum number of follicles ovulate at 500 IU (Wildt and Seager 1978). In ferrets, 4 µg of GnRH will produce LH release in excess of that induced by copulation (Carroll et al. 1985). Doses of 2 µg GnRH will induce LH release and induce ovulation in mink (Murphy 1979b). As GnRH has a much briefer persistence than hCG in serum, it may be a more useful substance for induction of ovulation. GnRH and GnRH analogues infused into the uterus of cattle or added to semen will induce large-scale release of LH from the pituitary (Manns et al. 1980). This finding suggests that GnRH in semen may be a potential method of delivery of the ovulatory stimulus to mustelids during artificial insemination.

Little information is available on the occurrence of embryonic mortality in mustelids. Withdrawal of ovarian or pituitary support will result in rapid embryonic death and resorption of fetuses in mink (Murphy et al. 1981; Murphy 1982). In the pseudopregnant ferret, augmentation of pituitary output of PRL and consequent elevation of luteal output of progesterone can be achieved by chronic treatment with the dopamine antagonist pimozide (Agu et al. 1986). This agent is orally active in the mink (Murphy, unpubl.) and can therefore be fed to mustelids to enhance the luteal phase secretion of progesterone. Further experimentation is necessary to establish whether more embryos can be induced to survive by this treatment.

It is known that the pineal principle, melatonin, secreted in a diurnal pattern, is responsible for annual reproductive cyclicity in many mammals, including the ferret (Baum et al. 1986). Although it has not been definitively established, most reports suggest that melatonin exerts its effects via its circulating concentration rather than through a direct neural pathway (Niles et al. 1977; Brown and Grota 1977). Therefore, a potential for immunoneutralization of melatonin via passive immunization exists. Use of monoclonal or polyclonal antibodies against melatonin may be a mechanism by which precocious puberty and ovarian recrudescence can be achieved in female mustelids.

References

Adams, C. E. 1976. Current research problems in the reproduction of the mink. *First International Congress of Fur Animal Production*, Helsinki, Finland. April 27–29.

Agu, G. O., et al. 1986. Evidence for the dopaminergic regulation and a luteotropic complex in the ferret. *Biol. Reprod.* 35:508–15.

Baum, M. J., and P. Schretlen. 1975. Neuroendocrine effects of perinatal androgenization in the male ferret. *Progr. Brain Res.* 42:343–55.

Baum, M. J., et al. 1986. Plasma and pineal melatonin levels in female ferrets housed under long or short photoperiods. *Biol. Reprod.* 34:96–100.

Brown, G. M., and L. J. Grota. 1977. Endocrine effects of the pineal gland and neutralization of circulating melatonin and N-acetylserotonin. *Can. J. Physio. Pharm.* 55:537–44.

Carroll, R. S., et al. 1985. Coital stimuli controlling luteinizing hormone secretion and ovulation in the female ferret. *Biol. Reprod.* 32:925–33.

Concannon, P. W., et al. 1980. Reflex LH release in estrous cats following single and multiple copulations. *Biol. Reprod.* 23:111–17.

Daniel, J. C. 1970. Coincidence of embryonic growth and uterine protein in the ferret. *J. Emb. Exp. Morph.* 24:305–12.

Deanesly, R. 1967. Experimental observations on the ferret corpus luteum of pregnancy. *J. Reprod. Fert.* 13:183–85.

Donovan, B. T. 1967. The control of the corpus luteum function in the ferret. *Arch. Anat. Microsc. Morph. Exp.* 56(Suppl. 3–4):315–25.

———. 1986. Is there a critical weight for oestrus in the ferret? *J. Reprod. Fert.* 76:491–97.

Donovan, B. T., and M. B. ter Haar. 1977. Effects of luteinizing hormone releasing hormone on plasma follicle-stimulating hormone and luteinizing hormone levels in the ferret. *J. Endocr.* 73:37–52.

Dufy-Barbe, L., et al. 1973. Time courses of LH and FSH release after mating in the female rabbit. *Endocr.* 92:1318–21.

Enders, R. K. 1952. Reproduction in mink. *Proc. Am. Phil. Soc.* 96:691–755.

Enders, A. C., and S. Schlafke. 1972. Implantation in the ferret: Epithelial penetration. *Am. J. Anat.* 133:291–316.

Enders, A. C., et al. 1986. Morphological changes in the blastocyst of the western spotted skunk during activation from delayed implantation. *Biol. Reprod.* 34:423–27.

Foresman, K. R., and R. A. Mead. 1978. Luteal control of nidation in the ferret (*Mustela putorius*). *Biol. Reprod.* 18:490–96.

Garcia-Mata, R. 1982. *El vison. Su cria en cautividad.* Buenos Aires: Hemisferio Sur.

Gonzalez-Mencio, F., et al. 1978. FSH and LH activity of PMSG from mares at different stages of gestation. *Anim. Reprod. Sci.* 1:137–44.

Greensides, R. D., and R. A. Mead. 1973. Ovulation in the spotted skunk (*Spilogale putorius latifrons*). *Biol. Reprod.* 8:576–84.

Hamilton, W. J. 1934. The early stages in the development of the ferret: Fertilization to the formation of the prochordal plate. *Trans. Roy. Soc. Edin.* 58:251–58.

Hammond, J., Jr. 1952. Gonadotrophin-induced ovulation in mink. *J. Mammal.* 33:218.

Hammond, J., Jr., and F. C. Chesterman. 1972. The ferret. In *The UFAW handbook on the care and management of laboratory animals,* 354–63. Edinburgh: E. and S. Livingston.

Hammond, J., and F. H. A. Marshall. 1930. Oestrus and pseudopregnancy in the ferret. *Proc. Roy. Soc. Lond.*, series B, 105:607.

Hammond, J., and A. Walton. 1934. Notes on ovulation and fertilization in the ferret. *Brit. J. Exper. Biol.* 2:307.

Hansson, A. 1947. The physiology of reproduction in mink (*Mustela vison*, Schreb.) with special reference to delayed implantation. *Acta Zool. Stockh.* 28:1–136.

Hartman, C. G. 1939. Ovulation, fertilization, and the transport and viability of eggs and spermatozoa. In *Sex and internal secretions,* ed. W. C. Young, 2d ed. Baltimore, Md.: Williams and Wilkins.

Heap, R. B., and J. Hammond, Jr. 1974. Plasma progesterone levels in pregnant and pseudopregnant ferret. *J. Reprod. Fert.* 39:149–52.

Hill, M., and A. S. Parkes. 1932. Studies on the hypophysectomized ferret. *Proc. Roy. Soc. Lond.*, series B, 112:146–58.

Hirshfield, A. N., and A. R. Midgely. 1978. The role of FSH in the selection of large ovarian follicles in the rat. *Biol. Reprod.* 19:606–11.

Kennedy, T. G. 1977. Evidence for a role for prostaglandins in the initiation of blastocyst implantation in the rat. *Biol. Reprod.* 16:286–91.

Kociba, G. J., and C. A. Caputo. 1981. Aplastic anemia associated with estrus in pet ferrets. *J. Am. Vet. Med. Assoc.* 173:1293–94.

Lindsell, C. E., et al. 1986. Variability in FSH:LH ratios among batches of commercially available gonadotrophins. *Theriogenol.* 25:167.

McPhail, M. K. 1935. Studies on the hypophysectomized ferret. IX. The effect of hypophysectomy on pregnancy and lactation. *Proc. Roy. Sci.*, series B., 117:34–47.

Manns, J. G., et al. 1980. Release of LH following intrauterine administration of gonadotrophin-releasing hormone. *Can. J. Anim. Sci.* 60:1023–26.

Marshall, F. H. A., and J. Hammond, Jr. 1945. Experimental control by hormone action of the oestrous cycle in the ferret. *J. Endocr.* 4:159–68.

Mead, R. A., and M. McRae. 1982. Is estrogen required for implantation in the ferret? *Biol. Reprod.* 27:540–47.

Mead, R. A., and A. Swannack. 1980. Aromatase activity in corpora lutea of the ferret. *Biol. Reprod.* 22:560–65.

Mead, R. A., et al. 1988. Changes in endometrial vascular permeability during the perimplantation period in the ferret (*Mustela putorius*). *J. Reprod. Fert.* 82:293–98.

Moller, D. M. 1973. The fine structure of lutein cells in the mink (*Mustela vison*) with special reference to the secretory activity during pregnancy. *Z. Zellforsch. Mikrosk. Anat.* 138:423–44.

Moon, Y. S., et al. 1976. Stimulating action of follicle stimulating hormone on estradiol 17-B stimulation by hypophysectomized rats in organ culture. *Endocr.* 97:244–47.

Murphy, B. D. 1979a. The role of prolactin in implantation and luteal maintenance in the ferret. *Biol. Reprod.* 21:517–21.

———. 1979b. Effects of GnRH on plasma LH and fertility in mink. *Can. J. Anim. Sci.* 59:25–33.

———. 1982. Similarities, differences in reproduction of mink, fitch. *Fur Rancher* 62:14–20.

———. 1986. Folliculogenesis and superovulation. In *Bovine embryo transfer course and workshop*, ed. R. J. Mapletoft, 18–27. Saskatoon, Saskatchewan: Univ. of Saskatchewan.

Murphy, B. D., P. W. Concannon, and H. F. Travis. 1982. Effects of medroxyprogesterone acetate on gestation in mink. *J. Reprod. Fert.* 66:491–97.

Murphy, B. D., and D. A. James. 1974. The effects of light and sympathetic innervation to the head on nidation in mink. *J. Exp. Zool.* 187:267–76.

Murphy, B. D., and R. A. Mead. 1976. Effects of antibodies to oestrogens on implantation in ferrets. *J. Reprod. Fert.* 46:261–63.

Murphy, B. D., and W. H. Moger. 1977. Progestins of mink gestation: The effects of hypophysectomy. *Endocr. Res. Comm.* 4:45–60.

Murphy, B. D., et al. 1980. Luteal function in mink: The effects of hypophysectomy after the preimplantation rise in progesterone. *Anim. Reprod. Sci.* 2:225–32.

Murphy, B. D., et al. 1981. Prolactin: The hypophyseal factor that terminates embryonic diapause in mink. *Biol. Reprod.* 25:487–91.

Murphy, B. D., et al. 1983. Luteal contribution to the termination of preimplantation delay in mink. *Biol. Reprod.* 28:497–503.

Murphy, B. D., et al. 1984. Variability in gonadotrophin preparations as a factor in the superovulatory response. *Theriogen.* 21:117–25.

Murphy, B. D., et al. 1987. Use of equine chorionic gonadotrophin in female mink during the breeding season. *Theriogen.* 28:667–74.

Niles, L. P., et al. 1977. Endocrine effects of the pineal gland and neutralization of circulating melatonin and N-acetylserotonin on plasma prolactin levels. *Neuroendocr.* 23:14–22.

Papke, R. L. 1980. Control of luteal function and implantation in mink by prolactin. *J. Anim. Sci.* 50:102–7.

Rider, V., and R. B. Heap. 1986. Heterologous antiprogesterone monoclonal
antibody arrests early embryonic development and implantation in the ferret
(*Mustela putorius*). *J. Reprod. Fert.* 76:459–70.

Robinson, A. 1918. The formation, rupture, and closure of ovarian follicles in
ferrets and ferret-polecat hybrids, and some associated phenomena. *Trans.
Roy. Soc. Edinburgh* 52:303–62.

Ryland, L. M. 1982. Remission of estrus-associated anemia following
ovariohysterectomy and multiple blood transfusions in a ferret. *J. Am. Vet.
Med. Assoc.* 81:820–22.

Ryland, L. M., and J. R. Gorham. 1978. The ferret and its diseases. *J. Am. Vet.
Med. Assoc.* 173:1154–58.

Sinha, A. A., and R. A. Mead. 1976. Morphological changes in the trophoblast,
uterus and corpus luteum during delayed implantation and implantation in
the western spotted skunk. *Am. J. Anat.* 145:331–56.

Travis, H. F., and P. E. Pilbeam. 1980. Use of artificial light and day length to
alter life cycles in mink. *J. Anim. Sci.* 50:1108–12.

Travis, H. F., et al. 1978. Relationship of vulvar swelling to estrus in mink. *J.
Anim. Sci.* 46:219–23.

Tritt, S. H. et al. 1986. Regulation of luteinizing hormone (LH) secretion in the
estrous ferret. *Biol. Reprod.* 34(Suppl.1):198.

Venge, O. 1959. Reproduction in the ferret and mink. *Anim. Breed. Abst.* 27:129–
45.

Wildt, D. E., and W. J. Seager. 1978. Ovarian response in the estrual cat receiving varying doses of HCG. *Hormone Res.* 9:144–50.

Wu, J. T., and M. C. Chang. 1972. Effects of progesterone and estrogen on the
fate of blastocysts in ovariectomized pregnant ferrets: A preliminary study.
Biol. Reprod. 7:231–37.

———. 1973. Hormonal requirements for implantation and embryonic development in the ferret. *Biol. Reprod.* 9:350–55.

9

RODNEY A. MEAD

Reproduction in Mustelids

Reproductive cycles of 19 species of mustelids have been reasonably well described, and fragmentary information regarding reproduction has been accumulated for another 19 of the existing 64 species. Reproductive patterns within the family are extremely diverse. Breeding times vary considerably between species so that one or more species may be breeding during any month of the year (Tables 9.1–9.3). The breeding season is relatively short in most females that produce a single litter each year, lasting only 1 to 2 months, but is prolonged in species such as *Mustela putorius* and *M. nivalis* that can produce 2 or more litters yearly. Some species have a single estrous cycle each breeding season, that is, they are monoestrous (Tables 9.1–9.3). Others exhibit recurring estrous cycles throughout the breeding season and are classified as polyestrous. The ferret (*M. putorius*) is considered polyestrous but remains in constant estrus for up to 5 months if not bred. The estrous cycle of unmated mustelids has been studied in only a few species. Consequently, our knowledge regarding the duration of estrus is limited, and some species that are now thought to be monoestrous may subsequently be classified as polyestrous. Estrus only lasts for 5 days in the European mink (*M. lutreola*) but will recur up to 3 times if the female is not bred (Moshonkin 1983). On the other hand, estrus usually lasts 8 to 9 days in the black-footed ferret (*M. nigripes*) and does not recur (Hillman and Carpenter 1983). Several species, such as *M. eversmanni*, *M. lutreola*, and *M. altaica*, that have only a single litter each year can be induced to produce second litters by removing the young shortly after birth and placing them with foster mothers (Tumanov 1977; Moshonkin 1981). A second estrus may also occur in monoestrous species if the female resorbs her embryos or loses her first litter within a few days after birth. Many but not all species of otters are polyestrous and have relatively short periods of estrus,

This work was partially funded by a grant from the National Institutes of Child Health and Human Development (HD06556).

lasting 3 to 5 days, that recur at approximately monthly intervals. The giant otter (*Pteronura*) and American river otter (*Lutra canadensis*) are both believed to be monoestrous, with heat lasting 14 days in the former species (Harrison 1963) and approximately 5 to 8 weeks in the latter (Stenson 1985, 1988). However, estrus may recur in *Pteronura* if the female loses her litter (Autuori and Deutsch 1977).

Mustelids also exhibit considerable diversity as to the age at which they first attain sexual maturity (Table 9.4). Females of some species such as *M. frenata* and *M. erminea* attain sexual maturity within 2 to 3 months after birth, whereas males of these same species do not reach puberty until they are yearlings. Only 30% of the female badgers (*Taxidea*) breed as juveniles (Messick 1981). The American pine marten (*Martes americana*) and the river otter (*Lutra canadensis*) do not breed until they are 24 to 27 months old. Stenson (1985) reports that only 55% of the wild female river otters in British Columbia bred when they were 3 years of age, whereas 90% of the older females were pregnant. Males of most species are not capable of breeding until they are about 1 year old. The western spotted skunk (*Spilogale gracilis*) is an exception in that most but not all males attain sexual maturity within 4½ to 5 months after birth.

The reproductive life span of mustelids has been well documented only for *M. vison* (Hansson 1947) and *M. mephitis* (Wade-Smith and Richmond 1975). Both species attain maximum reproductive potential at 2 years of age. Thereafter litter size steadily declines; however, mink still produce litters up to the age of 7 years (Enders, 1952). Carpenter (1985) suggested that reproduction may have failed for age-related reasons in a 7-year-old female black-footed ferret. Moshonkin (1983) reported normal reproduction in captive European mink (*M. lutreola*) up to the age of 5 to 6 years.

Three extremely different types of reproductive cycles can be distinguished within female mustelids. The primitive pattern of reproduction, from which the 2 other types of reproductive cycle are presumably derived, usually consists of a late winter or spring breeding period followed by a relatively short gestation of constant duration (Table 9.1). None of the species currently included in this group are known to exhibit a delay of blastocyst implantation; however, insufficient evidence is available to totally exclude its occurrence in many of these species. Blastocysts implant by day 11 in the least weasel (*M. nivalis*), making this the shortest preimplantation period thus far recorded in mustelids (Heidt 1970). Many species exhibiting short gestations with no delay of implantation have relatively restricted breeding seasons and give birth to a single litter each year. A few species in this group, such as the eastern spotted skunk (*Spilogale putorius*), pygmy spotted skunk (*S. pygmaea*), the

Table 9.1. Reproductive Characteristics of Species Exhibiting Short Gestation Periods with No Known Period of Delayed Implantation

Species	Distribution	Breeding Period	Type of Estrus[a]	Gestation	Litter Size[b]	Parturition	Source
I. Subfamily Lutrinae							
Aonyx capensis							
African clawless otter	Africa	Variable	—	63 days	2–5	Variable	Kingdon 1977
							Rosevear 1974
Aonyx cinerea							
Oriental small-clawed otter	Asia	Variable	P	60–64 days	1–6	Variable	Leslie 1970
							Timmis 1971
							Duplaix-Hall 1975
Pteronura brasiliensis							
Flat-tailed or giant otter	S. Amer.	July–Aug.	M	65–70 days	1–5	Aug.–Oct.	Autuori & Deutsch 1977
							Trebbau 1972
Lutrogale perspicillata							
Indian smooth-coated otter	Asia	Aug.	P	60–62 days	—	Oct.	Desai 1974
		Variable				Variable	Duplaix-Hall 1975
Lutra lutra							
European otter	Europe	Feb.–April	P	60–62 days	2–5	April–May	Novikov 1956
	Asia						Duplaix-Hall 1975
Lutra maculicollis							
Spotted-necked otter	Africa	July	—	60 days	2–3	Sept.	Procter 1963
II. Subfamily Mephitinae							
Conepatus mesoleucus							
Hog-nosed skunk	N. Amer.	Feb.	—	2 months	2–4	Apr.–May	Patton 1974
	S. Amer.						
Spilogale putorius							
Eastern spotted skunk	N. Amer.	Mar.–July	P	45–55 days	5	May–Aug.	Mead 1968a
Spilogale pygmaea							
Pygmy spotted skunk	Mexico	Mar.–June	—	48 days	2–6	May–Aug.	Teska et al. 1981

III. Subfamily Mustelinae

Species	Distribution	Mating season	Type[a]	Gestation	Litter size[b]	Birth season	References
Eira barbara / Tayra	Central & S. Amer.	April	M	63–67 days	3	June–July	Poglayen-Neuwall 1978; Encke 1968
Ictonyx striatus / Zorilla or striped polecat	Africa	Aug.–Nov.	P	35–44 days	1–3	Sept.–Dec.	Ball 1978; Rowe-Rowe 1978
Poecilictis libyca / N. African striped weasel	Africa	Feb.–May	P	37 days	1–3	Mar.–June	Petter 1959; Rosevear 1974
Poecilogale albinucha / African striped weasel	Africa	Aug.–Mar.	P	31–33 days	1–3	Sept.–Apr.	Rowe-Rowe 1978
Mustela nivalis / Least weasel	N. Amer. Europe	Spring Summer	P	34–37 days	1–7 (4.7)	Spring Summer Fall	Deanesly 1944; Heidt 1970
Mustela putorius / Ferret	Europe & Asia	Mar.–July	P	40–42 days	2–12	May–Aug.	Robinson 1918; Hammond & Walton 1934
Mustela altaica / Mountain weasel	Asia	Feb.–Mar.	M	40 days	7–8	Apr.–May	Novikov 1956; Roberts 1977
Mustela sibirica / Kolinsky mink	Asia	Feb.–Apr.(?)	P	34 days	2–10	Apr.–May(?)	Tumanov 1977; Novikov 1956
Mustela lutreola / European mink	Europe	Apr.	P	40–43 days	2–7 (2.3)	Apr.–May	Tumanov 1977
Mustela eversmanni / Steppe polecat	Europe & Asia	Mar.–Apr.	M	36–41 days	3–17 (8)	Apr.–May	Moshonkin 1981; Moshonkin 1983; Schmidt 1932
Mustela nigripes / Black-footed ferret	N. Amer.	Mar.–Apr.	M	42–45 days	2–6 (4.0)	May	Stroganov 1962; Hillman & Carpenter 1983; Thorne (in letters) 1987

[a]M = monoestrous, P = polyestrous
[b]Average litter size, if known, is given in parentheses.

Table 9.2. Reproductive Characteristics of Species Exhibiting Short Gestation Periods with Variable Periods of Delayed Implantation

Species	Distribution	Breeding Period	Type of Estrus[a]	Gestation (days)	Duration of Postimplantation	Litter Size[b]	Parturition	Source
I. Subfamily Mustelinae								
Mustela vison Mink	N. Amer. Europe Asia	Mar.–Apr.	P	40–75	28–30	1–17 (4.4)	Apr.–May	Enders 1952 Hansson 1947
II. Subfamily Mephitinae								
Mephitis mephitis Striped skunk	N. Amer.	Feb.–Apr.	M	59–77	—	1–10 (4.3)	May–June	Wade-Smith & Richmond 1975 Wade-Smith et al. 1980

[a]M = monoestrous, P = polyestrous
[b]Average litter size, if known, is given in parentheses.

Table 9.3. Reproductive Characteristics of Species Exhibiting Prolonged Gestations Accompanied by Delayed Implantation

Species	Distribution	Breeding Period	Type of Estrus[a]	Gestation	Duration of Post-implantation	Litter Size[b]	Parturition	Source
I. Subfamily Enhydrinae								
Enhydra lutris								
Sea otter	Pacific Ocean, N. Amer., USSR	Variable	P	6–7 months	—	1–2 (1)	Variable	Brosseau et al. 1975; Sinha et al. 1966
II. Subfamily Lutrinae								
Lutra canadensis								
American river otter	N. Amer.	Mar.–Apr.	M	245–365 days	—	1–3 (2)	Mar.–Apr.	Hamilton & Eadie 1964; Stenson, 1985; 1988
III. Subfamily Mustelinae								
Mustela Frenata								
Long-tailed weasel	N. Amer. S. Amer.	July	M	9 months	23–24	6–9 (6.8)	Apr.–May	Wright 1942; Wright 1963
Mustela erminea								
Short-tailed weasel or stoat	N. Amer. Europe	May–July	M	10–11 months	—	4–13 (6.1)	Apr.–May	Watzka 1940; Wright 1963
Gulo gulo								
Wolverine	USSR N. Amer.	May–July	—	8–9 months	—	3–4	Feb.–Apr.	Wright & Rausch 1955; Rausch & Pearson 1972
Vormela peregusna								
Marbled polecat	Europe Asia	Apr.–June	—	8–11 months	—	1–8	Jan.–Apr.	Mendelssohn et al. 1988
Martes pennanti								
Fisher	N. Amer.	Mar.–Apr.	—	327–358 days	—	3–4	Mar.–Apr.	Eadie & Hamilton 1958; Wright & Coulter 1967
Martes americana								
American marten	N. Amer.	July–Aug.	—	259–276 days	25–28	2–5	Mar.–Apr.	Pearson & Enders 1944; Wright 1963
Martes martes								
European pine marten	Europe	July	M	230–270 days 8–9 months	—	3–8	Mar.–Apr.	Canivenc et al. 1969; Canivenc 1970
Martes foina								
Stone or Beech marten	Europe	July	M	236–274 days 8–9 months	—	1–8	Mar.–Apr.	Canivenc et al. 1981; Novikov 1956
Martes zibellina								
Sable	Asia	June–July	P	253–297 days	—	1–5	Apr.–May	Novikov 1956; Bernatskii et al. 1976
Martes flavigula								
Yellow throated marten	Asia	Oct.–Nov.	—	172–190 days	—	1–5	Mar.–Apr.	Roberts 1977; Andriuskevicius 1982 (in letters)

(continued)

Table 9.3. (continued)

Species	Distribution	Breeding Period	Type of Estrus[a]	Gestation	Duration of Post-implantation	Litter Size[b]	Parturition	Source
IV. Subfamily Melinae								
Meles meles European badger	Europe Asia	Feb.–Mar.	P	345–365 days	45 days	2–6	Feb.	Neal & Harrison 1958 Canivenc & Bonnin 1981
Taxidea taxus American badger	N. Amer.	July–Aug.	M	7–8.5 months	—	2–3	Mar.–Apr.	Wright 1966
Arctonyx collaris Hog badger	Asia	Apr.–Sept.	P	5–9.5 months	—	2–4	Feb.	Parker 1979
V. Subfamily Mephitinae								
Spilogale gracilis Western spotted skunk	N. Amer.	Sept.–Oct.	P	210–260 days	28–31 days	1–6 (3)	Apr.–June	Mead 1968b Mead 1981

[a]M = monoestrous, P = polyestrous
[b]The average litter size, if known, is given in parentheses.

Table 9.4.   Age at Which Mustelids Are First Capable of Breeding

Species	Male (months)	Female (months)	Source
Mustela nivalis	8	4	Heidt 1970
Mustela erminea	13	2–2.5	Deanesly 1934, 1943
Mustela frenata	13	3	Wright 1947, 1963
Mustela putorius	10–11	7–12	Ishida 1968; Thorpe 1976
Mustela vison	10	10	Onstad 1967; Hansson 1947
Mustela lutreola	—	9–10	Moshonkin 1983
Mustela eversmanni	10	10	Novikov 1956
Spilogale putorius	11	11	Mead 1968a
Spilogale gracilis	4.5–5	4.5–5	Mead 1968b
Mephitis mephitis	10	10	Verts 1956
Martes martes	15	15	Novikov 1956
Martes foina	—	15	Canivenc et al. 1981
Martes americana	15	27–28	Jonkel & Weckworth 1963
Martes pennanti	12	12	Wright & Coulter 1967
Martes zibellina	15–16	15–16	Novikov 1956
Meles meles	24	12–16	Harrison 1963
Taxidea taxus	14	4–5	Wright 1966, 1969
Lutra canadensis	23–24	24–27	Hamilton & Eadie 1964
Gulo gulo	14–15	16–28	Rausch & Pearson 1972

least weasel (*M. nivalis*), zorilla (*Ictonyx*), and ferret (*M. putorius*) have extended breeding seasons and can have 2 or more litters a year (Gates 1937; Hartman 1964; Ball 1978; Teska et al. 1981). The latitude at which the animals live may in part influence whether or not a second litter is produced. For example, Mead (1968a) found no evidence that 2 litters could be produced by the eastern spotted skunk that inhabits Iowa and South Dakota (*S. putorius interrupta*) or in females of this subspecies that were captured in Texas and housed in Montana. On the other hand, Van Gelder (1959, 260) reported evidence for an extended breeding season and possibly 2 litters per year in the 2 southeastern subspecies *S. p. putorius* and *p. ambarvalis*. Litters of the least weasel (*M. nivalis*) were born throughout the year in captivity under nearly natural photoperiodic conditions (Heidt 1970), and Pohl (1910) reported finding pregnant females and young throughout the year in Germany.

   A second pattern of reproduction, consisting of relatively short but variable gestation periods, is known to occur only in the mink (*M. vison*) and striped skunk (*Mephitis mephitis*). Both species breed from late February to early April. The variability in gestation, which ranges from 40 to 75 days in the mink and from 59 to 77 days in the striped skunk, is due to the occurrence of a short period of delayed implantation, the duration of which is determined by the date of mating. Females bred early in the mating season have the longest pregnancies whereas those bred near the end of the season have short gestations with little or no delay of implantation. Ovulation in both species is induced by copulation and both remain receptive to the male until bred. Unlike the mink, the female striped skunk becomes very aggressive toward the male a day or 2 after breeding and will not allow the male near her (Wade-Smith and Richmond 1975), whereas the mink is routinely rebred 6 to 10 days after the first mating. This results in a second crop of follicles being ovulated and the expulsion from the uterus of most or all embryos from the first mating (Shackelford 1952). This type of reproductive cycle, characterized by short but highly variable gestation periods, may be more prevalent than is currently recognized and can only be detected by extensive captive-breeding programs.

The third pattern of female reproduction, which is known to occur in 16 species belonging to 10 different genera, is characterized by long gestation periods accompanied by prolonged periods of embryonic diapause and delayed implantation (Table 9.3). The embryos enter the uterus within 6 to 8 days after coitus, by which time they have undergone blastulation. Embryonic development then becomes arrested. The blastocysts remain in this state, referred to as embryonic diapause, for varying periods (Table 9.3), ranging from 5 months in the hog badger, *Arctonyx* (Parker 1979), to almost 1 year in the European badger, *Meles* (Canivenc and Bonnin 1981). Synchronization of renewed embryonic development and time of blastocyst implantation are controlled by changes in photoperiod (Mead 1981; Canivenc and Bonnin 1981). Postimplantation embryonic development is rapidly completed in as little as 23 to 24 days in the long-tailed weasel, *M. frenata* (Wright 1963), 25 to 28 days in *Martes americana* (Jonkel and Weckwerth 1963), 28 to 31 days in *Spilogale gracilis* (Foresmán and Mead 1973) and 45 days in *Meles* (Canivenc and Bonnin 1981).

Reproductive cycles of male mustelids have received considerably less attention than those of females; however, 2 distinct types of testicular cycles reportedly exist (Audy 1976). Males of most species studied thus far do not attain sexual maturity until they are at least 8 to 10 months old (Table 9.4), and sperm production is restricted to a relatively well-defined period lasting 3 to 4 months (Fig. 9.1). The testes and epididymides contain sperm long before females are in heat, whereas testicular regression usually begins before the breeding season ends. Consequently, few males may be capable of fertilizing females that lose their litters and exhibit a second estrus, as is the case with the striped skunk, *Mephitis* (Wade-Smith and Richmond 1978). The female western spot-

Testicular Cycle of North American Mustelids

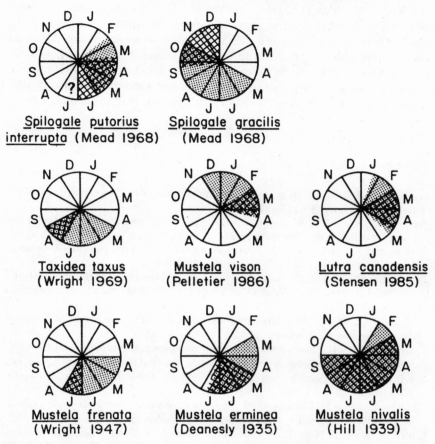

Figure 9.1. Stippled areas indicate the months during which sperm are found in testes of 8 North American mustelids. Crosshatched regions represent the time of estrus in females. Names in parentheses indicate the source of data for the testicular cycles.

ted skunk (*Spilogale gracilis*) is polyestrous and if not bred will continue to exhibit estrous cycles through December (Greensides and Mead 1973); however, testicular regression begins in early October. Although spermatogenesis in young spotted skunks lags behind that of adults by several months (sperm first found in testes of adults in April versus September for immature males), limited evidence suggests that testicular regression occurs simultaneously in both age groups (Mead 1968b). Consequently, all females must be bred by the end of October since males possess aspermatogenic testes in November.

The second pattern of reproduction in male mustelids is exhibited by the European badger (*Meles*). Sperm are found in the testes and epididymides throughout the year, however; the number of germ cells, testicular weight, Leydig cell size, and testosterone levels are distinctly higher immediately prior to the breeding season but then begin to decline somewhat before the breeding season is over (Audy 1976). Since sea otters breed throughout the year, the male sea otter may also exhibit continuous spermatogenesis; however, this supposition remains to be verified.

References

Audy, M. C. 1976. Le cycle sexuel saisonnier du mâle des mustélidés européens. *Gen. Comp. Endoc.* 30:117–27.

Autuori, M. P., and L. A. Deutsch. 1977. Contribution to the knowledge of the giant Brazilian otter *Pteronura brasiliensis*. *Zool. Garten.* 47:1–8.

Ball, M. P. 1978. Reproduction in captive-born zorillas at the National Zoological Park, Washington. *Intern. Zoo Yearbook* 18:140–43.

Bernatskii, V. G., et al. 1976. Natural and induced ovulation in the sable (*Martes zibellina* L.). *Dokl. Akad. Nauk SSSR* 230:1238–39. Translated by Consultants Bureau, New York.

Brosseau, C., et al. 1975. Breeding the sea otter (*Enhydra lutris*) at Tacoma Aquarium. *Intern. Zoo Yearbook* 15:144–47.

Canivenc, R. 1970. Photopériodisme chez quelques mammifères à nidation différée. In *La photorégulation de la reproduction chez les oiseaux et les mammifères*, ed. J. Benoit and I. Assenmacher, 453–66. Paris: Centre National de la Recherche Scientifique.

Canivenc, R., and M. Bonnin. 1981. Environmental control of delayed implantation in the European badger (*Meles meles*). *J. Reprod. Fert. Suppl.* 29:25–33.

Canivenc, R., et al. 1969. Induction de nouvelles générations lutéales pendant la progestation chez la Martre européene (*Martes martes* L.). *C. R. Acad. Sci. Paris*, séries D, 269:1437–40.

Canivenc, R., et al. 1981. Delayed implantation in the beech marten (*Martes foina*). *J. Zool.* 193:325–32.

Carpenter, J. W. 1985. Captive breeding and management of black-footed ferrets. *Black-footed Ferret Workshop Proceedings*, 12.1–12.13. Cheyenne, Wyo.: Wyoming Game and Fish Publ.

Deanesly, R. 1935. The reproductive processes of certain mammals. IX. Growth and reproduction in the stoat (*Mustela erminea*). *Phil. Trans. Roy. Soc. London*, series B, 225:459–92.

———. 1943. Delayed implantation in the stoat (*Mustela mustela*). *Nature* 151:365–66.

———. The reproductive cycle of the female weasel (*Mustela nivalis*). *Proc. Zool. Soc. London* 114:339–49.

Desai, J. H. 1974. Observations on the breeding habits of the Indian smooth otter. (*Lutrogale perspicillata*) in captivity. *Intern. Zoo Yearbook* 14:123–24.

Duplaix-Hall, N. 1975. River otters in captivity: A review. In *Breeding endangered species in captivity*, ed. R. D. Martin, 315–27. New York: Academic Press.

Eadie, W. R., and W. J. Hamilton, Jr. 1958. Reproduction in the fisher in New York. *N.Y. Fish and Game J.* 5:77–83.

Encke, W. 1968. A note on the breeding and rearing of tayras (*Eira barbara*) at Krefeld Zoo. *Intern. Zoo Yearbook* 8:132.

Enders, R. K. 1952. Reproduction in the mink (*Mustela vison*). *Proc. Amer. Philos. Soc.* 96:691–755.

Foresman, K. R., and R. A. Mead. 1973. Duration of post-implantation in a western subspecies of the spotted skunk (*Spilogale putorius*). *J. Mammal.* 54:521–23.

Gates, W. H. 1937. Spotted skunks and bobcat. *J. Mammal.* 18:240.

Greensides, R. D., and R. A. Mead. 1973. Ovulation in the spotted skunk (*Spilogale putorius latifrons*). *Biol. Reprod.* 8:576–84.

Hamilton, W. J., Jr. and W. R. Eadie. 1964. Reproduction in the otter, *Lutra canadensis*. *J. Mammal.* 45:242–52.

Hammond, J., and A. Walton. 1934. Notes on ovulation and fertilization in the ferret. *J. Exp. Biol.* 11:307–19.

Hansson, A. 1947. The physiology of reproduction in mink (*Mustela vison* Schreb.) with special reference to delayed implantation. *Acta Zool.* 28:1–136.

Harrison, R. J. 1963. A comparison of factors involved in delayed implantation in badgers and seals in Great Britain. In *Delayed Implantation*, ed. A. C. Enders, 99–114. Chicago: Univ. of Chicago Press.

Hartman, L. 1964. The behaviour and breeding of captive weasels (*Mustela nivalis* L.). *New Zealand J. Sci.* 7:147–56.

Heidt, G. A. 1970. The least weasel, *Mustela nivalis* Linneaus. Developmental biology in comparison with other North American *Mustela*. *Pub. Mus. Michigan State Univ., Biol. Series* 4(7):227–82.

Hill, M. 1939. The reproductive cycle of the male weasel (*Mustela nivalis*). *Proc. Zool. Soc. London*, series B, 109:481–512.

Hillman, C. N., and J. W. Carpenter. 1983. Breeding biology and behavior of captive black-footed ferrets (*Mustela nigripes*). *Intern. Zoo Yearbook* 23:186–91.

Ishida, K. 1968. Age and seasonal changes in the testis of the ferret. *Arch. Histol. Jap.* 29:193–205.

Jonkel, C. J., and R. P. Weckwerth. 1963. Sexual maturity and implantation of blastocysts in the wild pine marten. *J. Wildl. Mgmt.* 27:93–98.

Kingdon, J. 1977. *East African mammals*. Vol. 3, Part A (Carnivores). New York: Academic Press.

Leslie, G. 1970. Observations on the Oriental short-clawed otter (*Aonyx cinerea*) at Aberdeen Zoo. *Intern. Zoo Yearbook* 10:79–81.

Mead, R. A. 1968a. Reproduction in eastern forms of the spotted skunk (Genus *Spilogale*). *J. Zool., London* 156:119–36.

———. 1968b. Reproduction in western forms of the spotted skunk (Genus *Spilogale*). *J. Mammal.* 49:373–90.

————. 1981. Delayed implantation in the mustelidae with special emphasis on the spotted skunk. *J. Reprod. Fert.* (Suppl.) 29:11–24.

Mendelssohn, H., M. Ben-David, and S. Hellwing. 1988. Reproduction and growth of the marbled polecat (*Vormela peregusna syriaca*) in Israel. *J. Reprod. Fert.* Abstract Series 1:20.

Moshonkin, N. N. 1981. Potencialnaya poliestrichnost evropeiskoi norki (*Lutreola lutreola*) (Potential polyestricity of the mink [*Lutreola lutreola*]). *Zool. Zh.* 60:1731–34.

————. 1983. Reproduktivnii cikl samok evropeiskoi norki (*Lutreola lutreola*) (The reproductive cycle in females of the European mink [*Lutreola lutreola*]). *Zool. Zh.* 62:1879–83.

Neal, E. G., and R. J. Harrison. 1958. Reproduction in the European badger (*Meles meles* L.). *Trans. Zool. Soc. London* 29:67–131.

Novikov, G. A. 1962 [1956]. *Carnivorous mammals of the fauna of the USSR.* Zoological Institute of the Academy of Science of the USSR, no. 62. Translated by Israeli Program for Scientific Translations, Jerusalem: Israeli Program for Scientific Translations.

Onstad, O. 1967. Studies on postnatal testicular changes, semen quality, and anomalies of reproductive organs in mink. *Acta Endocr. Copenh.* 55 (Suppl.) 117:1–117.

Parker, C. 1979. Birth, care and development of Chinese hog badgers (*Arctonyx collaris albogularis*) at Metro Toronto Zoo. *Intern. Zoo Yearbook* 19:182–85.

Patton, T. S. 1974. Ecological and behavioral relationships of the skunks of Tans-Pecos Texas. Ph.D. diss., Texas A and M Univ., College Station.

Pearson, O. P., and R. K. Enders. 1944. Duration of pregnancy in certain mustelids. *J. Exp. Zool.* 95:21–35.

Petter, F. 1959. Reproduction en captivité du zorille du Sahara (*Poecilictis libyca*). *Mammalia* 23:378–80.

Poglayen-Neuwall, I. 1978. Breeding, rearing and notes on the behavior of tayras (*Eira barbara*). *Intern. Zoo Yearbook* 18:134–40.

Pohl, L. 1910. Wieselstudien. *Zool. Beob.* 51:234.

Procter, J. 1963. A contribution to the natural history of the spotted-necked otter (*Lutra maculicollis* Lichtenstein) in Tanganyika. *E. Afr. Wildl. J.* 1:93–102.

Rausch, R. A., and A. M. Pearson. 1972. Notes on the wolverine in Alaska and the Yukon Territory. *J. Wildl. Mgmt.* 36:249–68.

Roberts, T. J. 1977. *The mammals of Pakistan.* London: Ernst Benn Ltd.

Robinson, A. 1918. The formation, rupture and closure of ovarian follicles in ferrets and ferret-polecat hybrids and some associated phenomena. *Trans. Roy. Soc. Edin.* 52:303–62.

Rosevear, D. R. 1974. The carnivores of West Africa. Pub. No. 723. London: British Mus. Nat. Hist.

Rowe-Rowe, D. T. 1978. Reproduction and post-natal development of South African mustelines (*Carnivora: Mustelidae*). *Zool. Africana* 13:103–14.

Schmidt, F. 1932. Der Steppeniltis (*Putorius eversmanni* Less.). *Deutsche Pelztierzuchter* 7:453–58.

Shackelford, R. M. 1952. Superfetation in the ranch mink. *Am. Nat.* 86:311–19.

Sinha, A. A., et al. 1966. Reproduction in the female sea otter. *J. Wildl. Mgmt.* 30:121–30.

Stenson, G. B. 1985. The reproductive cycle of river otters. Ph.D. diss., Univ. of British Columbia, Vancouver.

Stenson, G. B. 1988. Oestrus and the vaginal smear cycle of the river otter, *Lutra canadensis. J. Reprod. Fert.* 83:605–10.

Teska, W. R., et al. 1981. Reproduction and development of the pygmy spotted skunk (*Spilogale pygmaea*). *Am. Midland Nat.* 105:390–92.

Thorpe, D. H. 1967. Basic parameters in the reaction of ferrets to light. *Ciba Found. Study Group* 26:53–70.

Timmis, W. H. 1971. Observations on breeding the Oriental short-clawed otter (*Amblonyx cinerea*) at Chester Zoo. *Intern. Zoo Yearbook* 11:109–11.

Trebbau, P. 1972. Notes on the Brazilian giant otter (*Pteronura brasiliensis*) in captivity. *Zool. Garten.* 41:152–56.

Tumanov, I. L. 1977. O potentsial'noi poliestrichnosti nekotorykh vidov Kun'ikh (Mustelidae) (On potential polyestrus in some species of the Mustelidae). *Zool. Zh.* 56:619–25.

Van Gelder, R. G. 1959. A taxonomic revision of the spotted skunks (Genus *Spilogale*). *Bull. Amer. Mus. Nat. Hist.* 117:233–392.

Verts, B. J. 1967. *The biology of the striped skunk.* Chicago: Univ. of Illinois Press.

Wade-Smith, J., and M. E. Richmond. 1975. Care, management, and biology of captive striped skunks (*Mephitis mephitis*). *Lab. Animal Sci.* 25:575–84.

———. 1978. Reproduction in captive striped skunks (*Mephitis mephitis*). *Am. Midland Nat.* 100:452–55.

Wade-Smith, J., et al. 1980. Hormonal and gestational evidence for delayed implantation in the striped skunk, (*Mephitis mephitis*). *Gen. Comp. Endocr.* 42:509–15.

Watzka, M. 1940. Mikroskopisch-anatomische Untersuchungen über die Ranzzeit und Tragdauer des Hermelins (*Putorius ermineus*) (Microscopic-anatomical investigations concerning the breeding season and gestation period of the ermine [*Putorius ermineus*]). *Ztschr. Mikro.-anat. Forschung.* 48:359–74.

Wright, P. L. 1942. Delayed implantation in the long-tailed weasel (*Mustela frenata*), the short-tailed weasel (*Mustela cicognani*), and the marten (*Martes americana*). *Anat. Rec.* 83:341–53.

———. 1947. The sexual cycle of the male long-tailed weasel (*Mustela frenata*). *J. Mammal.* 28:343–52.

———. 1963. Variations in reproductive cycles in North American mustelids. In *Delayed implantation,* ed. A. C. Enders, 77–97. Chicago: Univ. of Chicago Press.

———. 1966. Observations on the reproductive cycle of the American badger (*Taxidea taxus*). In *Comparative biology of reproduction in mammals,* ed. I. W. Rowlands, 27–45. New York: Academic Press.

———. 1967. Reproduction and growth in Maine fishers. *J. Wildl. Mgmt.* 31:70–87.

———. 1969. The reproductive cycle of the male American badger (*Taxidea taxus*). *J. Reprod. Fert.* (Suppl.) 6:435–45.

Wright, P. L., and R. Rausch. 1955. Reproduction in the wolverine (*Gulo gulo*). *J. Mammal.* 36:346–55.

10

JOE HERBERT

Light as a Multiple Control System on Reproduction in Mustelids

Successful reproduction is primarily a matter of timing, particularly that of the birth season and the interval between litters. In most species, births occur during the spring and early summer (Table 10.1), when the food supply is plentiful and variations in air temperature are less extreme. Mechanisms are also needed, however, to restrict reproduction by spacing births. Parturition, lactation, and maternal behavior are all very costly to a female, in terms of both energy requirements and the effects they have on social behavior (Trivers 1985): for example, females may be required to defend the young during the neonatal period. Males also pay behavioral penalties for breeding: because intrasexual competition is a feature of sexual selection, males become more aggressive toward one another during the breeding season, with the attendant risk of injury. Finally, there are ecological restraints: too many growing young may put too great a demand on resources; in addition, undue strains on the social system of the species may develop as the young become increasingly independent and then become pubertal and reproductively competent themselves. These are the ultimate factors controlling the timing of reproduction.

The neuroendocrine system of the animal uses cues known as *proximate* factors to time components of the reproductive cycle; alterations in day length are the major (though not the only) proximate factor in many species, including mustelids (Herbert 1977). The mustelids offer some particularly good examples of the way in which changes in day length affect reproductive activity and of how this can differ between species. Mustelids, like other genera, presumably use the cue

I am grateful to my colleagues Michael Hastings, Tarvinder Juss, and Andrea Walker for improving this manuscript, to Jane Rowell for preparing it for publication, and to Karen Gangel for excellent editing. The work of our laboratory is supported by grants from the Medical Research Council.

Table 10.1. Classification of Mustelid Breeding Cycle According to Phase of Annual Photoperiod Operative during Components of the Female's Reproductive Activity (See Also Fig. 10.1)

Species	Taxonomic Name	Male Mating Season	Female Mating Season	Ovul.	Imp. Season	Birth Season
Type 1						
Ferret	M. furo	Jan.–Sept.	Mar.–Sept.	Cop.	None	May–Oct.
Eastern spotted skunk	S. putorius	Mar.–May	Mar.–May	Spon.	None	Apr.–May
Type 2						
Mink	M. vison	Dec.–Apr.	Feb.–Mar.	Cop.	Apr.	Apr.–May
Striped skunk	M. mephitis	Feb.–Apr.	Feb.–Apr.	Cop.	Mar.–Apr.	May–June
Type 3						
American badger	T. taxus	July–Oct.	July–Oct.	?	Dec.–Feb.	Mar.–Apr.
Western spotted skunk	S. putorius	Aug.–Oct.	Sept.–Oct.	Cop.	Apr.–May	May–June
Type 4						
Badger	M. meles	Jan.–Mar.	Feb.–Mar.	Spon.	Dec.–Jan.	Jan.–Feb.
Pine marten	M. martes	May–July	June–July	?	Mar.	Apr.–May
Stoat	M. erminea	May–July	May–July	Spon.	Feb.–Mar.	Mar.–Apr.
Stone marten	M. foina	Apr.–July	May–July	?	Feb.–Mar.	May

Note: Ovul. = ovulation; cop. = copulation-induced; spon. = spontaneous; imp. = implantation.

provided by day length because it is a highly reliable indicator of the passage of seasons. Measuring day length, under most conditions, allows the animal to predict the likely climatic conditions over the next few months and thus to prepare for a birth season still to come (Sadleir, 1969).

The 3 basic components of the mustelid reproductive cycle are mating, diapause (which does not occur in some species), and gestation. The female, as the center of each of these processes, is the focus of several control systems (Fig. 10.1). In the male, however, the mating season is the only reproductive factor that can be controlled by light.

Day length can not act directly on the time of birth because this is more or less fixed for a given species by developmental necessities. Day length therefore acts on events that time implantation, not on birth itself. In some species, the time between mating and birth can be made more flexible by delaying implantation for a variable period, during which the embryo remains dormant (diapause). In such cases, there is a second level of control. The duration of diapause is also regulated by the annual light cycle (Pearson and Enders 1944; Hansson 1947; Mead 1971). Both of these controls regulate the phase of the birth season relative to the time of year. In addition, the duration of the breeding season is regulated by day length; this third control limits the maximal amount of reproductive activity in a season.

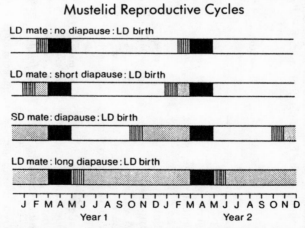

Figure 10.1. Four annual breeding cycles of female mustelids. The sequence of the mating (represented by lines), diapause (dots), and birth (black) seasons are shown over a 2-year period. The birth seasons of each type have been aligned for diagrammatic purposes; specific examples of the timing of seasonal reproductive events are given in Table 10.1. (See text for further explanation.)

The Breeding Season

All mustelids for which adequate data exist appear to breed seasonally, with the possible exceptions of the sea otter and the European otter in some localities (Chanin 1985). It seems likely that the time of the year during which males are in season, and the duration of the season itself, is tailored to fit the reproductive requirements of the female. The males of many mustelid species (and other taxa) seem to begin their season before the females (Table 10.1). There may be good biological reasons for this: males often have to initiate breeding activity, or take part in preliminary behavior (such as obtaining territories), which forms an essential part of the mating process. However, defining the beginning and end of the male's cycle is more difficult than defining that of the female, and this may account (at least in part) for the apparent difference in the timing of the breeding season between the sexes. Nevertheless, there may be real sex differences in the phase relation between photoperiod duration and the mating season—for example, mink (Boissin-Agasse et al. 1982) and badger (Audy 1976a)—though the subject has been studied very little so far.

There are 4 basic types of annual cycle in female mustelids. In the simplest form (type 1 in Table 10.1 and Fig. 10.1), mating is followed by fertilization and implantation, without an intervening period of diapause, as in the ferret (Mead and Wright 1983). In such species, the mating season usually begins in late winter or early spring—that is, when day length is increasing. Similarly, type 2 cycles are characterized by very short periods of embryonic diapause preceding implantation, so that mating and births occur during the same half of the year (as in type 1). Type 3 cycles differ because the mating season occurs in the autumn or early winter, as day length is decreasing, and is followed by a longer period of diapause and finally birth the following spring. In type 4 cycles, spring mating is followed by an even longer diapause, so that the interval between mating and birth is almost 1 year; but the light conditions during the mating season are similar to those of types 1 and 2. The distinctions between the different types of cycle are thus: whether diapause occurs; the duration of day length during the mating season; and whether mating and births occur during the same or subsequent years, that is, whether there is a long or a short period of diapause (see also Mead and Wright 1983).

During the breeding (mating) season, the male usually continues to be reproductively active, but the female has 2 strategies (Eckstein and Zuckerman 1956; Schwartz 1973). In such species as the ferret, mink, and striped skunk, the female ovulates in response to coitus. Ovarian discharge of ova thus occurs only when fertilization is likely. Females of other species (the badger, spotted skunk, and stoat) ovulate spontaneously (that is, even in the absence of a male). In rats, which also ovulate spontaneously, ovulation is controlled by light and occurs during the first part of the night when the animals become behaviorally active. This maximizes fecundity by linking the times of estrous behavior and ovulation via light (Hastings and Herbert 1986). It would be interesting to know if the time of ovulation can be controlled by light in female mustelids that ovulate spontaneously, some of which are nocturnal (western spotted skunk and badger) and others diurnal (polecat and stoat).

Light Systems Controlling Reproduction

It is customary to divide mustelids and other seasonal species into "long-day" or "short-day" breeders. This distinction was initially based on the onset of the mating season under natural conditions, and subsequently on the type of experimental photoperiod that caused premature activation of the gonads in reproductively quiescent animals. It seems, however, that dividing species in this way is no longer useful. The time of breeding relative to day length may occur either before or after the winter solstice in the same species depending on latitude (Lloyd and Englund 1973). This clearly shows that long- or short-day breeding is mutable and not an invariant property of a species. Secondly, all species so far examined need alternating periods of lengthening and shortening day lengths if the annual cycle is to occur normally. This need occurs regardless of whether the animal begins its season of reproductive activity when day lengths are increasing or decreasing. A phenomenon used in the study of circadian rhythms, however, provides a more useful analogy than the long- or short-day classifications. During the 24 hours of the day-night cycle, a number of events (eating, ovulation, changes in levels of corticoids in the blood, and so on) occur in sequence, each entrained to the time of day. This sequence depends upon each event having its own temporal relation to the phase of the circadian cycle— that is, its "phase angle." A more satisfactory way of understanding the relation between the timing of mating seasons and day length is, therefore, to postulate a similar phase angle between components of the annual breeding cycle and the annual light cycle. The phase angle is modifiable by (1) properties inherent in the neuroendocrine system of a given species (including the way in which components of its cycle respond to light), (2) sex differences within a species in this system, (3) an animal's previous photoperiodic experience, and (4) other environmental factors (both physical and biological) that may "gate" or modify the effect of photoperiods. The reasons for given circadian events having particular phase angles have been extensively discussed (see Hastings and Herbert 1986). The phase angle is very sensitive to changes in the period of the endogenous circadian oscillator. Whether the same properties control the annual phase angle is not clear, because it has been hard to demonstrate endogenous circannual cycles in, for example, ferrets kept in constant lighting conditions (Thorpe 1967).

Figure 10.2 shows that the phase angle is sensitive to the period of the entraining rhythm: the shorter this period, the larger the phase angle, so that ferrets being driven by 4-month cycles actually appear to breed during short photoperiods. An animal's gonadal response to a given photoperiod depends not only on the duration of the photoperiod, but on its immediately previous photoperiodic history. The solstices are

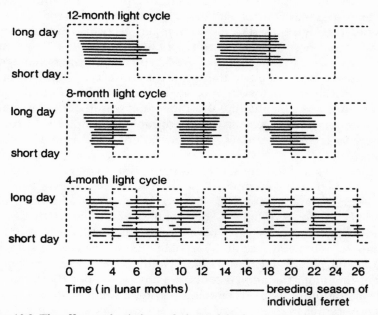

Figure 10.2. The effect on the timing and phase of the female ferret's breeding (estrous) season of exposure to alternating sequences of long (14L:10D) or short (8L:16D) photoperiods. Each line represents the occurrence of a period of reproductive activity. Three conditions are shown: in the top section, ferrets were exposed to alternating photoperiods of 6 months each; in the middle 4 months; and below, 2 months. Each condition changes the period, duration, and phase (relative to the photoperiod) of successive breeding seasons.

therefore important under natural conditions in enabling an animal to synchronize its reproduction with the light cycle.

Photorefractoriness

However, another way of controlling reproduction is known: photorefractoriness, or insensitivity to a photoperiod. Ferrets kept in long photoperiods resembling midsummer eventually go out of estrus (though later than normal) and will not return to breeding condition unless the photoperiods are first shortened and then lengthened again (Donovan 1976). The current idea is that a photoperiod inducing a given reproductive state also sets up a condition of developing refractoriness to itself, thus limiting its duration of action and causing the animal's cycle to enter the alternative (or "default") mode. Photorefractoriness is being recognized as a most important control mechanism in the regulation by light of fertile and infertile parts of the annual cycle. Refractoriness may control both onset and duration of breeding. Hamsters persistently exposed to short photoperiods (which initially inhibit reproduction) eventually cease to respond to them and so begin their breeding season; and the timing of the sheep's breeding season may owe as much to refractoriness to the suppressive effects of long (summer) days as to the stimulatory role of short ones (Robinson et al. 1985). Though it is plausible that the short breeding season of the mink (and other mustelids such as stoats) is due to rapid development of refractoriness to lengthening days, definite experimental evidence is still lacking.

Under normal conditions, refractoriness is overcome by the animal being exposed to the photoperiod opposite to which it is refractory (Fig. 10.3). For example, ferrets that have become refractory to long photoperiods (16 hours of light [L] and 8 hours of darkness [D]), and are therefore anestrous, are resensitized to the stimulatory effects of long photoperiods by being put into short photoperiods (8L:16D) for about 7 weeks (Thorpe and Herbert 1976a). Nothing overt seems to happen during this time; the animal continues anestrous. But there must be some alteration in its neuroendocrine system, for reexposure to long days now results in regrowth of the gonads and a new breeding season. During the autumn, therefore, it seems likely that such species as mink and ferrets are resensitized to the lengthening days of the next spring by the short days of the previous winter. Converse effects during spring follow if any part of the animal's cycle has become refractory to short photoperiods (Fig. 10.3). Refractoriness may also occur in young ani-

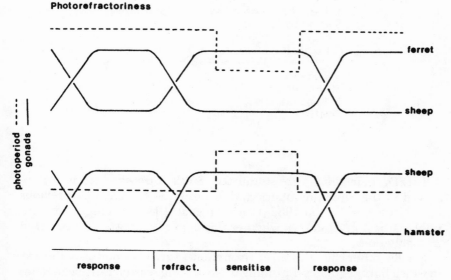

Figure 10.3. Diagram of the relation between photoperiod (dotted line) and development and dispersion of photorefractoriness. The upper part shows photorefractoriness to long photoperiods, which induce gonadal activity in some species (ferret), but inactivity in others (sheep). Gonadal activity in such species during prolonged exposure to long photoperiods, then short ones, and finally reexposure to long photoperiods is shown by the solid line. The lower part of the diagram shows the converse effects in species becoming photorefractory to short days. Response: animal's neuroendocrine system responding to the prevailing photoperiod; Refract.: system refractory to prevailing photoperiod; Sensitize: system being resensitized by exposure to opposite photoperiod; Response: system responds again to original photoperiod.

mals before their first breeding season. Ferrets born into, and left in, long photoperiods have their first season much later than normal, showing that a period of short days is needed to time puberty normally (Thorpe 1967).

The importance of an animal's photoperiodic history is evident in the way it measures the photoperiod and defines a long or short day. These parameters seem not to be absolute. Clearly, if mustelids are to use photoperiod as a predictor of subsequent events, they must be able to distinguish, for example, autumnal from vernal equinox, even though the absolute photoperiod of both is the same. If hamsters are exposed for several weeks to very long photoperiods of 18L:6D (during which their gonads develop) and are then transferred to a photoperiod of 12L:12D, they treat the new light regime as short, and their gonads regress. If another group is first exposed for the same period to 6L:18D (during which their gonads regress) and is then moved to the same

Table 10.2. Variations in Badger Birth
Season by Latitude

Locality	Time of birth
SW France	End Jan.
SE England	Early Feb.
N England	End Feb.
Scotland	Early Mar.
Russia	End Mar.

Source: Neal 1986.

12L:12D, their gonads are stimulated. So the opposite effects are observed in the same photoperiod, depending on the animal's previous light history (Robinson 1985; Hoffman et al. 1986; Hastings et al., unpubl.). It is not yet known whether this applies to mustelids, though it probably does.

Finally, other climatic and environmental variables modulate the way in which light regulates the timing of the mating season. European foxes are distributed widely throughout Europe, but breed progressively later with increasing latitude (Lloyd and Englund 1973). Similar evidence (Table 10.2) exists for the spotted skunks of North America and the badgers of Europe (Mead 1968; Neal 1986). The critical second factor has not been identified: in the case of latitude, it is not easy to invoke relative readings of the same photoperiod, since animals at higher latitudes experience a shorter day during the winter than those nearer the equator and should, on this basis, respond more rapidly to the subsequently lengthening days—the converse of what is actually observed.

Light and the Reproductive Cycle

One might assume that increasing day length controls the onset of breeding of types 1, 2, and 4; and there is considerable experimental evidence to support this. One of the earliest demonstrations of the reproductive effects of artificial light, repeatedly confirmed by others, was made on the ferret, when Bissonnette (1932) showed that extra light given to anestrous females during the winter provoked premature breeding (Fig. 10.4). Similar results have been described for the mink and stone marten (Duby and Travis 1972; Audy 1976b).

Experiments on species other than mustelids argue caution: this is not the only possible explanation for light-controlled spring breeding. The spring breeding season of hamsters, for example, is regulated by the

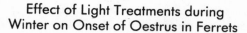

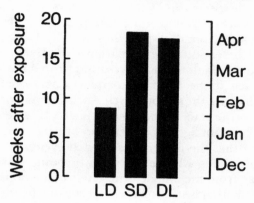

LD: long photoperiods (14L:10D)
SD: short photoperiods (8L:16D)
DL: daylight

Figure 10.4. The effect on the onset of gonadal activity of exposing female ferrets near the winter solstice to either long photoperiods (LD, 14L:10D), short photoperiods (SD, 8L:16D), or natural daylight (DL). The histograms show the time taken for estrus to appear after the start of each treatment.

short days of the preceding autumn, which drive the animal out of breeding condition. Breeding starts at the appointed time because hamsters eventually become photorefractory to short days, thus allowing the reproductive system to escape the suppressive effects of short days (Reiter 1983). Ferrets and hamsters both have spring breeding seasons: but the active control in ferrets seems to be lengthening days that time the onset of breeding; whereas in hamsters, shortening photoperiods time the onset of anestrus and hence, secondarily, the breeding season. It is important to recognize that for successful reproductive timing mechanisms determining an animal's nonmating period (offset) are as significant as those controlling onset.

The proximate control of the breeding season of mustelids that mate in the autumn may, by analogy with species such as the sheep (which do likewise), be a response to shortening photoperiods (Karsch et al. 1984). Sheep exposed to unseasonal short days (8L:16D) show premature breeding; male mink treated similarly do likewise (Boissin-Agasse et al. 1982). Whether other autumn-breeding mustelids, such as the western spotted skunk, might respond to similar regimes is not yet known.

Breeding may not occur simply as an immediate response to the stimulus of the appropriate photoperiod. If this were the case, then preventing a ferret from experiencing lengthening days would prevent (or at least delay) the start of the breeding season. But ferrets kept in midwinter photoperiods from the winter solstice (Fig. 10.4) still come into season at about the expected time (Carter et al. 1982). Some process seemingly causes a default timing mechanism to operate, but this effect can be overridden if the animal experiences abnormally long photoperiods or changes in the seasonal cycle, such as those following movement from the northern to southern hemisphere. If ferrets are kept for several years in constant light, however, they breed irregularly (Thorpe 1967), indicating that the light cue is needed eventually and that the mechanism determining whether or not a breeding season will occur may be separable from the one determining when it occurs.

Delayed implantation is said to be a characteristic feature of mustelid reproduction, though not all species exhibit it (Eckstein and Zuckerman 1956). The duration of true pregnancy (the time between implantation and birth) is approximately within the same range for all mustelids (though longer in larger forms such as the badger than in smaller ones such as the weasel). Since most species give birth in the spring and summer, implantation occurs during and is controlled by the period of lengthening days. Nidation is accelerated by artificially long photoperiods in mink and the western spotted skunk (Fig. 10.5) (Pearson and Enders 1944; Mead 1971). In species whose spring mating is followed by a short period of diapause (type 2), implantation occurs when the photoperiod reaches a given value (Allais and Martinet 1978; Wade-

Diapause in Western Spotted Skunk

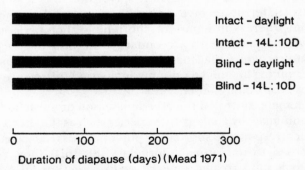

Figure 10.5. The duration of diapause in both intact and blind western spotted skunks exposed to artificially long photoperiods (14L:10D) during early spring compared with those kept in daylight (drawn from data in Mead 1971).

Smith et al. 1980). The badger, which has a longer period of true gestation, may implant in December and thus be subject to different light control; exposing badgers to short photoperiods during the summer induced premature implantation (Canivenc and Bonnin 1975), though in this experiment temperature was also lowered, making the role of light difficult to distinguish. Latitude may also have an effect on the time of implantation in this species, since diapause ends later in more northerly regions (Neal 1986). The photoperiodic control of diapause thus focuses the time of births into a very restricted period, offering another accurate control of the mating season.

Species with long periods of diapause (with an interval between mating and birth of nearly a year) need another mechanism to assure that their reproduction remains locked to the annual light cycle. Lactation in many species is characterized by infertility; but some species, particularly small rodents that need to maximize reproduction, show the phenomenon of postpartum estrus. The female mates and becomes pregnant again shortly after giving birth, so that another litter is ready to be born as soon as the previous one is weaned. Mustelids such as the badger, stoat, and marten also show postpartum estrus (Eckstein and Zuckerman 1956)—in their case because there would otherwise be insufficient time for fertilization to occur during the annual cycle. In such species, therefore, the timing of diapause (by regulating the time of births) also determines the onset of the breeding season.

Diapause may also function to limit reproduction. Species showing diapause have, as a rule, only 1 litter per year, whereas those without delayed implantation may have several.

The duration of the breeding season is also controlled by light, but this differs according to species. The breeding season of the ferret (like that of the European weasel) lasts until late summer or early autumn, though individual differences in the time of offset are markedly greater than for onset (Herbert and Vincent 1972). Shortening day lengths could be responsible for ending the season. Experimental findings support this; the season of estrous female ferrets is curtailed by exposing them to prematurely short days (Thorpe and Herbert 1976a). The breeding season of species such as the mink, however, is very short, lasting not much more than a month. At its end, therefore, day lengths are still increasing, so that this control process is not available for terminating breeding. The possible role of photorefractoriness in limiting short seasons has already been discussed.

Animals do not need to experience the total duration of the photoperiod to diagnose long or short photoperiods. For example, anestrous ferrets exposed to a single 7-hour period of light do not alter the timing of their breeding, but if they are given two 3½-hour periods separated by a 5-hour dark interval (a "skeleton" photoperiod), they respond as if they had been exposed to a single long light period of about 14 hours (Hammond 1951; Hart 1951). (Similar findings have been made in sheep and other species.) This ability has obvious biological importance: it allows the animal to sample its environment and arrive at an accurate estimate of the photoperiod. Fossorial or crepuscular species, in particular, need this ability. One interpretation of these findings is that the initial period of light (simulating dawn) sets up a subsequent interval of light sensitivity; if the animal experiences light during any part of this interval—which it might well do under conditions of long days—then its neuroendocrine system behaves as if the illumination between the 2 periods of light had been constant. Testicular growth in mink can be stimulated in this way (Boissin-Agasse et al. 1982). This mechanism is sometimes referred to as external coincidence since it postulates coincidence between an external event (light) and some internal state (altered sensitivity to illumination).

Detecting the Light Signal: The Pineal

The neuroendocrine system of mustelids responds to photoperiods by altering gonadotropin output from the pituitary. Increasing the pulses of luteinizing hormone (LH) and follicle stimulating hormone (FSH) stimulates the gonads to produce both mature gametes and steroid hormones; decreasing pulses brings breeding to a close (Baum 1986). Less is known about the endocrine changes responsible for the end of diapause (Mead 1968; Martinet and Allain 1985). But how is the duration of a photoperiod translated into an endocrine output signal?

The receptors for the light stimulus lie in the eye. Blind ferrets do not respond to light, but show intermittent periods of estrus unrelated to the seasons (Herbert et al. 1978). Recent neuroanatomical evidence suggests that the pathways from the eye enter the hypothalamic suprachiasmatic nucleus in ferrets and spotted skunks as well as other mammals (Thorpe 1975; Thorpe and Herbert 1976b; Moore 1983; May et al. 1985). This nucleus is crucial for the expression of circadian rhythms and their entrainment or synchronization with the daily light-dark cycle. There are connections from the nucleus to another adjacent structure, the paraventricular nucleus, which in turn sends nerve fibers down the spinal cord to the cells which project to the superior cervical nucleus and thence to the pineal (Moore 1983).

Pinealectomized ferrets no longer respond either to experimental long photoperiods in the spring, which would otherwise induce early breeding, or to premature short ones in the summer, ordinarily bringing breeding to an early close (Fig. 10.6) (Herbert 1969, 1981; Thorpe and Herbert 1976a). The same result follows removal of the superior cervical ganglia, which supply the pineal with its secretomotor nerve supply. Similar findings are known for other seasonal light-sensitive species, such as the sheep, vole, and hamster (Bittman 1984; Hastings and Herbert 1986). Pinealectomized animals seem totally unresponsive to experimental alteration of photoperiod.

Removal of the pineal from ferrets does not, however, abolish intermittent periods of breeding. Pinealectomized females show recurrent periods of estrous activity, but henceforth unrelated to the time of year (Herbert et al. 1978). Furthermore, animals pinealectomized during the spring and kept in either natural or lengthened photoperiods still begin their season that year at the same time as normal ferrets exposed only to daylight. In subsequent years, estrus occurs late and is unsynchronized between animals (Fig. 10.7) (Herbert 1971, 1981; Herbert et al. 1978). The animal therefore needs the pineal (and changing photoperiods) to

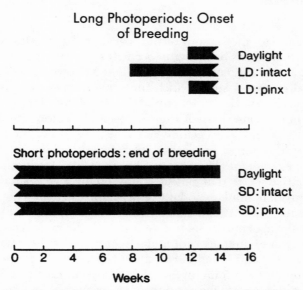

Figure 10.6. The effect of pinealectomy on the onset (above) and the duration (below) of the breeding season in ferrets. Pinealectomy prevents the acceleration of gonadal activity that otherwise follows exposure to long photoperiods in winter (above) and the premature termination of activity after exposure to short photoperiods in the summer (below). In both cases, the animals' reproductive systems behave as if they had not been exposed to the experimental light regime.

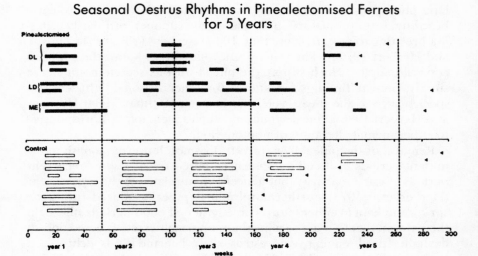

Figure 10.7. Recurrent periods of estrus over several years in either pinealectomized (above) or normal ferrets (below) kept in daylight. Each series of bars is the estrous record of an individual animal.

synchronize it with the seasons of the year, but also employs a fail-safe mechanism that allows breeding to occur for 1 year at the correct time even in the absence of control (Herbert and Klinowska 1978).

The third control system, which times nidation in species showing diapause, may also be dependent upon the pineal. Blind western spotted skunks do not implant early if they are exposed to long photoperiods (Mead 1971). Would pinealectomy have the same effect? Removing the superior cervical ganglia prevents mink from implanting early after exposure to extra light (Murphy and James 1974; Martinet and Allain 1985), which suggests that pinealectomy might give a similar result. Melatonin given to mink exposed to extra light also delays implantation (Martinet and Allain 1985). But pinealectomy of pregnant skunks or mink (or removal of autonomic nerve supply to the pineal) has no effect on implantation of animals kept in daylight (Mead 1972; Martinet and Allain 1985; May and Mead 1986): an obvious parallel to the effect of pinealectomy on the timing of estrus in ferrets in normal daylight. In both cases, a default (or fail-safe) mechanism may be operating. However, melatonin given either as an evening injection or as an implant delays implantation under similar conditions, as does blinding (May and Mead 1986); but these results would not be predicted based on results in ferrets. It seems that the exact function of the pineal in timing diapause in mustelids must still be worked out.

The pineal is activated by darkness, rather than light; that is, its nerve supply stimulates it to secrete during the night. The nature of this secretion, and the way it changes brain function, is critical to our understanding of how the central nervous system translates a light signal into an endocrine output.

Although the pineal is metabolically very active, and a great number of potentially neuroactive substances are found within it, at the moment only one of them, melatonin, has an established role in light measurement. Melatonin is formed from its precursor, serotonin (5-hydroxytryptamine), by a 2-stage process (Klein 1985). Darkness, by stimulating the afferent nerves to the pineal, increases the amount of the first of the 2 enzymes that form it, and hence its concentration in the gland. This causes more melatonin to be secreted, and it is the form of this secretory pattern that seems to be the essential message telling the brain (and hence the pituitary) the duration of the night and thus the photoperiod. The longer the night, the longer the period of increased melatonin levels in the ferret (Fig. 10.8), as in several other species (Baum et al. 1986; Hastings et al. 1985c). Because the enzyme has to be newly formed as darkness falls, there is usually a lag before melatonin levels rise; but as soon as the animal experiences light, the formation of melatonin halts abruptly, though how this occurs is still not understood. A critical point is that, if for any reason the continuity of melatonin is interrupted, the system reading the melatonin signal defaults; and even if melatonin levels are quickly reinstated, the 2 periods of darkness are not read as 1.

Since darkness stimulates melatonin production and light inhibits it, the passage of the animal from one photoperiod to another will be reasonably accurately reflected in the pattern of melatonin secretion. Add to this the fact that melatonin must be produced in an unbroken stream in order to be read as a long night, and the way the photoperiod is sampled becomes clear. As soon as an animal samples light, its melatonin levels will fall precipitously, and the read-out system will reset (Hastings et al. 1985a).

The absence (or low levels) of melatonin during daylight also has a function: it seems to reset the readout mechanism so that the next elevation of melatonin is recognized as the start of the night. This may be one function of the suprachiasmatic nuclei in the context of seasonal breeding; by locking the melatonin rhythm to the daily light-dark cycle, they may enable its secretion to remain in phase with daily rhythms. If ani-

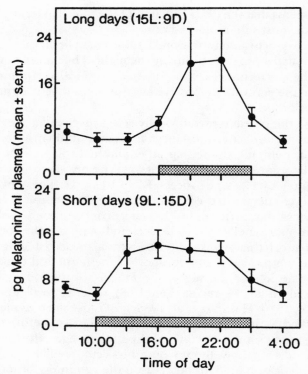

Figure 10.8. Daily rhythms of melatonin in the blood of ferrets kept in either long or short photoperiods. The nocturnal rise mirrors the duration of the dark phase (night) (Baum et al. 1986).

mals such as hamsters are implanted with capsules that release melatonin continuously, the readout system may become confused and fail to detect the changes in the animal's endogenous, pineal-derived melatonin secretion, and thus not respond to experimental photoperiods (Hastings et al. 1988).

More directly, however, injecting extra melatonin in the correct pattern can deceive an animal into responding as if it were in short days (long nights). Thus, melatonin antagonizes the effect of long photoperiods on anestrous ferrets, and treated animals come into estrus at the normal time, rather than prematurely (Herbert 1971). Estrous ferrets given melatonin also behave as if they were in short days and go out of estrus early (Thorpe and Herbert 1976a). Similar findings follow treatment of male stoats and mink with melatonin implants (Rust and Meyer 1969; Allain and Rougeot 1980). Melatonin can, therefore, alter both the timing and the duration of the breeding season in ferrets. It can also replicate a third, and biologically most important, action of short days in this species: it can disperse refractoriness. If ferrets refractory to long photoperiods are given melatonin for several weeks, upon cessation of treatment they respond to the long photoperiod by coming into estrus once again, just as if they had spent the treatment period in short days (Thorpe and Herbert 1976a).

So far, the converse has not been reported in mustelids: melatonin injected in a long day pattern has not been reported to replicate its effects (Carter et al. 1982). It is important to recognize that the gonadal response to melatonin can vary between species, just as their responses to the same photoperiod vary.

The way in which the brain reads the melatonin signal can also be varied. Short-day photorefractory hamsters do not respond to melatonin with the usual gonadal atrophy (Bittman 1984). Yet they can still detect changes in the melatonin pattern, since the dispersal of photorefractoriness depends upon the pineal being present. Furthermore (in sheep at least), the pattern of melatonin secretion in short days seems the same whether or not the animals are photosensitive (Karsch et al. 1986). Finally, there is recent evidence that hamsters changed to schedules of 14L:10D from either 18L:6D or 8L:16D, though their reproductive systems react in opposite ways, nevertheless have similar profiles of melatonin secretion (Hoffmann et al. 1986; Hastings et al., unpubl.). The system reading the melatonin signal can alter its properties, so that at some stage between the signal and the output (not necessarily the readout mechanism directly) there can be a change in the interpretation of the signal itself; and this depends upon the previous photoschedule to which the animal has been exposed. Gradually some of the biological effects are being explained in terms of the responsible neuroendocrine mechanisms, though reconciliation of the two still has a long way to go.

Which part of the central nervous system (CNS) reads the melatonin signal? Though destroying the hypothalamic suprachiasmatic nuclei (SCN) disturbs a hamster's ability to respond to a short photoperiod, injecting melatonin still causes gonadal regression (Bittman et al. 1979). So the readout mechanism remains intact despite removal of the SCN, which are more concerned with generating the signal. Large lesions of the anterior hypothalamus in out-of-season male or female ferrets cause rapid growth of the gonads relative to controls, even though the animals remain in short days (Donovan and van der Werff ten Bosch 1959; Baum and Goldfoot 1974). Implanting melatonin directly into the hypothalamus of deer mice resulted in their becoming anestrous (Glass and Lynch 1982). All these experiments point, therefore, to a site in the hypothalamus which might read the melatonin signal. It has recently been reported that lesions in this area that do not damage the SCN (and therefore leave the animal's circadian rhythms intact) prevent the male hamster from responding to short photoperiods (Hastings et al. 1985c). There is no recent evidence for mustelids, though it is highly likely that they will share a common CNS mechanism with rodents. Yet we still do not know, for any species, how the hypothalamus computes the duration of the melatonin signal, why it apparently resets if melatonin secretion is interrupted, or how it modulates gonadal response according to an animal's photoperiodic history or latitude. It is the brain that allows species or sexual differences to occur in the photoinduced control of reproduction, and the brain that enables an individual to adapt its response according to its circumstances. We therefore need to understand exactly how the central nervous system interprets the information it receives from the environment via the pineal.

References

Allain, D., and J. Rougeot. 1980. Induction of autumn moult in mink (*Mustela vison* Peale and Beavois) with melatonin. *Reprod. Nutr. Develop.* 20:197–201.

Allais, C., and L. Martinet. 1978. Relation between daylight ratio, plasma progesterone levels and timing of nidation in mink (*Mustela vison*). *J. Reprod. Fert.* 45:133–36.

Audy, M.-C. 1976a. Influence du photopériodisme sur la physiologie testiculaire de la fouine (*Martes foina erx.*) (The effect of photoperiods on testicular activity in stone martens). *C. R. hebd. Séanc. Acad. Sci. Paris sér. D* 283:805–8.

———. 1976b. Le cycle sexuel saisonnier du male des mustélidés européens (The seasonal sexual cycle of male European mustelids). *Gen. Comp. Endoc.* 30:117–27.

Baum, M., and D. A. Goldfoot. 1974. Effect of hypothalamic lesions on maturation and annual cyclicity of the ferret testis. *J. Endocr.* 62:59–73.

Baum, M. 1986. Use of the ferret in reproductive neuroendocrinology. In *Ferret Medicine*, ed. J. G. Fox and S. M. Niemi. Philadephia: Lea and Febiger.

Baum, M., C. A. Lynch, C. A. Gallagher, and M-H Deng. 1986. Plasma and pineal melatonin levels in female ferrets housed under long or short photoperiods. *Biol. Reprod.* 34:95–100.

Bedford, Duke of, and F. H. A. Marshall. 1842. On the incidence of the breeding season in mammals after transference to a new latitude. *Proc. R. Soc. Lond.,* series B, 130:396–99.

Bissonnette, T. H. 1932. Modification of mammalian sexual cycles: Reactions of ferrets (*Putorius vulgaris*) of both sexes to electric light added after dark in November and December. *Proc. R. Soc. Lond.,* series B, 110:322–36.

Bittman, E. L. 1984. Melatonin and photoperiodic time measurement: Evidence from rodents and ruminants. In *The pineal gland*, ed. R. J. Reiter, 155–92. New York: Raven Press.

Bittman, E. L., B. D. Goldman, and I. Zucker. 1979. Testicular responses to melatonin are altered by lesions of the suprachiasmatic nuclei in golden hamsters. *Biol. Reprod.* 21:647–56.

Boissin-Agasse, L., J. Boissin, and R. Ortavant. 1982. Circadian photosensitive phase and photoperiodic control of testis activity in the mink (*Mustela vison* Peale and Beavois), a short day mammal. *Biol. Reprod.* 26:110–19.

Canivenc, R., and M. Bonnin. 1975. Les facteurs écophysiologiques de régulation de la fonction lutéale chez les mammifères à ovo-implantation différée (Ecophysiological factors regulating luteal function in mammals with delayed implantation). *J. Physiol.* (Paris) 70:533–38.

Carter, D. S., J. Herbert, and P. M. Stacey. 1982. Modulation of gonadal activity by timed injections of melatonin in pinealectomised or intact ferrets kept under two photoperiods. *J. Endocr.* 93:211–22.

Chanin, P. 1985. *The natural history of otters*. London: Croom Helm.

Donovan, B. T. 1976. Light and the control of the oestrous cycle in the ferret. *J. Endocr.* 39:105–13.

Donovan, B. T., and J. J. van der Werff ten Bosch. 1959. The relationship of the hypothalamus to oestrus in the ferret. *J. Physiol.* 147:93–108.

Duby, R. T., and H. F. Travis. 1972. Photoperiodic control of fur growth and reproduction in the mink. *J. Exp. Zool.* 182:217–26.

Eckstein, P., and S. Zuckerman. 1956. The oestrous cycle in the mammalia. In *Marshall's physiology of reproduction*, vol. 1, 226–396. London: Longmans.

Glass, J. D., and G. R. Lynch. 1982. Evidence for a brain site of melatonin action in the white footed mouse, (*Peromyscus leucopus*). *Neuroendocr.* 34:1–6.

Hammond, J. 1951. Control by light of reproduction in ferrets and minks. *Nature* 167:150–51.

Hansson, A. 1947. The physiology of reproduction in mink with special reference to delayed implantation. *Acta Zool.* 28:1–136.

Hart, D. S. 1951. Photoperiodicity in the female ferret. *J. Exp. Zool.* 28:1–12.

Hastings, M. H., and J. Herbert. 1986. Endocrine rhythms. In *Neuroendocrinology*, ed. S. Lightman and B. J. Everitt. Oxford: Blackwells.

Hastings, M. H., J. Herbert, N. D. Martensz, and A. C. Roberts. 1985a. Annual reproductive rhythms in mammals: Mechanisms of light synchronization. *Ann. New York Acad. Sci.* 453:182–204.

———. 1985b. Melatonin and the brain in photoperiodic mammals. In *Photoperiodism, melatonin and the pineal*. Ciba Fndn. Symp. 117:57–77. London: Pitman.

Hastings, M. H., A. C. Roberts, and J. Herbert 1985c. Neurotoxic lesions of the anterior hypothalamus disrupt the photoperiodic but not the circadian system of the Syrian hamster. *Neuroendocr.* 40:316–24.

Hastings, M. H., A. P. Walker, A. C. Roberts, and J. Herbert. 1988. Intrahypothalamic melatonin blocks photoperiodic responsiveness in the male Syrian hamster. *Neuroscience* 24:987–91.

Herbert, J. 1969. The pineal gland and light-induced oestrus in ferrets. *J. Endocr.* 43:625–36.

———. 1971. The role of the pineal gland in the control by light and the reproductive cycle of the ferret. In *The pineal gland,* ed. G. W. Wolstenholme and J. Knight, 303–27. Ciba Fndn. Symp. London: Churchill.

———. 1977. External factors and ovarian activity in mammals. In *The ovary,* 2d ed., ed. P. Eckstein and S. Zuckerman, 485–505. New York: Academic Press.

———. 1981. The pineal gland and photoperiodic control of the ferret's reproductive cycle. In *Biological clocks in seasonal cycles,* ed. B. F. and D. E. Follet, 261–76. Bristol, U.K.: Wright.

Herbert, J., and M. Klinowska. 1978. Day length and the annual reproductive cycle in the ferret (*Mustela furo*): The role of the pineal body. In *Environmental endocrinology,* ed. I. Assenmacher and D. S. Farner, 87–93. Berlin: Springer.

Herbert, J., P. M. Stacey, and D. H. Thorpe. 1978. Recurrent breeding seasons in pinealectomised or optic-nerve sectioned ferrets. *J. Endocr.* 78:389–97.

Herbert, J., and D. S. Vincent. 1972. Light and the breeding season in mammals. *Intern. Cong. Series Endocrinology* 273:875–79. Amsterdam: Excerpta Medica.

Hoffmann, K., H. Illnerova, and J. Vanecek. 1986. Change in duration of the nighttime melatonin peak may be a signal driving photoperiodic responses in the Djungarian hamster (*Phodopus sungorus*). *Neuroscience Letters* 67:68–72.

Karsch, F. J., E. L. Bittman, D. L. Foster, R. L. Goodman, S. J. Legan, and J. E. Robinson. 1984. Neuroendocrine basis of seasonal reproduction. *Rec. Prog. Horm. Res.* 40:185–232.

Karsch, F. J., E. L. Bittman, J. E. Robinson, S. M. Yellon, N. L. Wayne, D. H. Olster, and A. D. Kaynard. 1986. Melatonin and photorefractoriness: Loss of response to the melatonin signal leads to seasonal reproductive transitions in the ewe. *Biol. Reprod.* 34:265–74.

Klein, D. C. 1985. Photoneural regulation of the mammalian pineal gland. In *Photoperiodism, melatonin and the pineal.* Ciba Fndn. Symp. 117:38–57. London: Pitman.

Lloyd, H. G., and J. Englund. 1973. The reproductive cycle of the red fox in Europe. *J. Reprod. Fert.* (Suppl.) 19:119–30.

Martinet, L., and D. Allain. 1985. Role of the pineal gland in the photoperiodic control of reproductive and nonreproductive functions in the mink (*Mustela vison*). In *Photoperiodism, melatonin and the pineal.* Ciba Fndn. Symp. 117:170–87. London: Pitman.

May, R., M. DeSantis, and R. A. Mead. 1985. The suprachiasmatic nucleus and retinohypothalamic tract in the western spotted skunk. *Brain Res.* 339:378–81.

May, R., and R. A. Mead. 1986. Evidence for pineal involvement in timing implantation in the western spotted skunk. *J. Pineal. Res.* 3:1–8.

Mead, R. A. 1968. Reproduction in western forms of the spotted skunk (genus *Spilogale*). *J. Mammal.* 49:373–90.

———. 1971. Effects of light and blinding upon delayed implantation in the spotted skunk. *J. Reprod. Fert.* 5:214–20.

————. 1972. Pineal gland: Its role in controlling delayed implantation in the spotted skunk. *J. Reprod. Fert.* 30:147–50.

Mead, R. A., and P. L. Wright. 1983. Reproductive cycles of mustelidae. *Acta Zool. Fennica* 174:169–72.

Moore, R. Y. 1983. Organization and function of a central nervous system oscillator: The suprachiasmatic nucleus. *Fed. Proc.* 42:2783–89.

Murphy, B. D., and D. A. James. 1974. The effects of light and sympathetic innervation to the head on nidation in mink. *J. Exp. Zool.* 187:267–76.

Neal, E. 1986. *The natural history of badgers.* London: Croom Helm.

Pearson, O. P., and R. K. Enders. 1944. Duration of pregnancy in certain mustelids. *J. Exp. Zool.* 95:21–35.

Reiter, R. J. 1983. Seasonal reproductive events related to the pineal gland. In *The pineal gland and its endocrine role,* ed. J. Axelrod, F. Franschini, and G. P. Vela, 303–16. New York: Plenum Press.

Robinson, J. E. 1985. The reproductive response of the ewe to day length depends on photoperiodic history. *Soc. Reprod. Ann. Meeting Abst.* 41.

Robinson, J. E., N. L. Wayne, and F. J. Karsch. 1985. Refractoriness to inhibitory day lengths initiates breeding season of the Suffolk ewe. *Biol. Reprod.* 32:1024–30.

Rust, C. C., and R. K. Meyer. 1969. Hair color, molt and testis size in male short-tailed weasels treated with melatonin. *Science* 165:921–22.

Sadleir, R. F. M. S. 1969. *The ecology of reproduction in wild and domestic mammals.* London: Methuen.

Schwartz, N. B. 1973. Mechanisms controlling ovulation in small mammals. In *Handbook of physiology,* sect. 7, Endocrinology, vol. 2, 125–41. Baltimore: Williams and Wilkins.

Thorpe, D. H. 1967. Basic parameters in the reactions of ferrets to light. *Ciba Fndn. Study Group* 26:33–66.

Thorpe, P. A. 1975. The presence of a retinohypothalamic projection in the ferret. *Brain Res.* 85:343–46.

Thorpe, P. A., and J. Herbert. 1976a. Studies on the duration of the breeding season and photorefractoriness in female ferrets pinealectomised or treated with melatonin. *J. Endocr.* 70:255–62.

Thorpe, P. A., and J. Herbert. 1976b. The accessory optic system of the ferret. *J. Comp. Neurol.* 170:295–310.

Trivers, R. 1985. *Social evolution.* Menlo Park, Calif.: Cummings.

Wade-Smith, J., M. E. Richmond, A. Mead, and H. Taylor. 1980. Hormonal and gestational evidence for delayed implantation in the striped skunk, (*Mephitis mephitis*). *Gen. Comp. Endocr.* 42:509–15.

11

DAVID E. WILDT AND KAREN L. GOODROWE

The Potential for Embryo Technology in the Black-Footed Ferret

Embryo technology provides the means for the rapid dispersal of genetic material of outstanding individuals. Theoretically, the techniques of embryo collection, storage, and transfer as well as the formation of embryos by *in vitro* fertilization may eventually have an impact on conservation and artificial breeding of rare and endangered species.

Embryo techniques that might be applied to the black-footed ferret are considered in this chapter. Such discussion requires preemptive warning: although conceptually available, the actual application of the technology to preservation of the black-footed ferret would not be a trivial endeavor. Embryo studies in the order Carnivora have been few and are limited essentially to the domestic dog (*Canis familiaris*), cat (*Felis catus*), mink (*Mustela vison*) and domestic ferret (*Mustela putorius*). Liveborn offspring have been reported only rarely from such research. The meager success rate, however, is more a reflection of the lack of opportunity and financial incentives than of biological limitations. The usefulness of embryo techniques for black-footed ferret propagation will be determined ultimately by the availability of 3 essential resources: (1) animals, (2) expertise, and (3) equipment and supplies.

Embryo transfer is the process by which preimplantation embryos are collected from one female (the donor) and transferred to other females (recipients) to complete gestation. In vitro fertilization (IVF) is the fusion of gametes (ova and spermatozoa) outside of the female's reproductive tract in a culture dish maintained in a controlled, artificial environment. Fertilized ova resulting from IVF then undergo embryo transfer

National Zoological Park research projects discussed in this report were funded by the Friends of the National Zoo (FONZ). Karen L. Goodrowe is supported, in part, by a grant from the Women's Committee of the Smithsonian Associates, Smithsonian Institution, Washington, D.C. The authors thank Patricia Schmidt, Lyndsay Phillips, and Mitchell Bush for their helpful comments and Stephanie Michie for preparing the manuscript.

into the reproductive tract of the original ovum donor (autotransfer) or into the uterus or oviduct of an unrelated recipient. After a species-specific interval, the 1-cell fertilized ovum cleaves into 2 cells (blastomeres), both of which continue to divide until the embryo is a compacted ball of approximately 32 cells (a morula). Further development results in blastulation or the formation of a blastocoele or cavity within the embryo. The embryo, now a blastocyst, contains an inner cell mass at one pole that eventually gives rise to the fetus. The remaining thin layer of cells surrounding the blastocoele is the trophoblast, which eventually forms the embryonic membranes. With further expansion in most species, the blastocyst hatches from the zona pellucida (the protective covering encompassing the embryo) and undergoes implantation. In the rabbit and ferret, however, implantation is accomplished without prior hatching. Because of the implications to organ transplantation, the terminology embryo transplant is rarely used in modern literature, embryo transfer being preferable.

Advantages of Embryo Technology

The benefits derived from embryo recovery and transfer and IVF for the commercial livestock industry and infertile human couples have been reviewed (Gwatkin 1977; Rogers 1978; Whittingham 1979; Blandau 1980; Wright and Bondioli 1981). For the black-footed ferret, a seasonal breeder, embryo recovery and transfer could potentially permit the collection of embryos repeatedly throughout the year. Recovered embryos could be transferred to another species of ferret serving as a surrogate both pre- and postnatally. Alternatively, using IVF, mature oocytes could be retrieved from the ovarian follicles and mixed with spermatozoa in a specialized medium, producing embryos in an artificially controlled environment. This system has the same major advantage as embryo recovery and transfer, because oocytes could be collected throughout the year and added to fresh or frozen-thawed spermatozoa; the embryos could then be transferred to the ferret recipient. With both procedures the reproductive competence of the valuable donor female is not wasted on gestation of the developing fetuses, but is expended on the repeated production of the genetic material necessary for formation of viable embryos. The major differences in benefits between the two approaches will become apparent, however. Whereas embryo recovery and transfer are technically simpler than IVF, the latter has the potential of being less stressful and dangerous to the donor. The IVF system also is not restricted by sexually incompatible mating pairs, since fertilization can occur in the absence of estrus or actual animal-to-animal contact.

The primary disadvantage of embryo recovery and transfer and IVF is complexity, since disruption of any one of numerous critical factors can undermine either system. General factors influencing embryo transfer in animals have been summarized (Adams 1982) and are reviewed here in relation to a possible ferret program.

The selection, management, and genotype of the donor can have considerable effect on reproductive performance in an embryo or IVF program. Satisfactory physical condition is important to the success of hormonal therapy, mating, and withstanding the stresses associated with anesthesia and with oocyte and embryo recovery. There is some controversy regarding the quality of oocytes maturing naturally versus those induced by the use of exogenous gonadotropins. Although some data indicate that these ova are of comparable quality, more recent information suggests a higher proportion of abnormal embryos in females hormonally treated (Fujimoto et al. 1974; Maudlin and Fraser 1977; Seidel 1981; Goodrowe et al. 1986). Likewise, there appears to be a greater incidence of unfertilized oocytes in at least one species inseminated with frozen-thawed rather than fresh spermatozoa (Newcomb 1980). One of the primary factors affecting embryo quality is the genotype of the donor. Studies in our laboratory demonstrate remarkable differences among mouse strains in embryo number, quality, and freezability after using a standardized gonadotropin treatment (Schmidt et al. 1985).

Selection of which oocytes should be designated for an IVF attempt or which embryos should be transferred is based primarily on morphological criteria. Such measures are subjective and not always infallible—not all high-quality, transferred embryos result in offspring; and, conversely, an embryo with questionable features may, on occasion, implant and develop normally. In general, the normal expected stage of development and morphological uniformity are reliable estimates of viability. However, considerable variation may exist in embryo morphology among species, particularly in symmetry of cell cleavage and size and appearance of individual blastomeres. Therefore, standards formulated in one species cannot necessarily be extrapolated to another species. Experience in assessing the oocytes and embryos of a particular species is essential and is complicated in carnivores, including the ferret, by the dark cytoplasm of the oocyte and embryo. This characteristic obscures the identification of the pronuclei, in the case of early-stage IVF oocytes, and individual blastomeres in more developed embryos. Whether this feature will inhibit the successful freeze-thawing of carnivore embryos, as observed in dark, lipid-containing pig embryos (Polge and Willadsen 1978) remains to be determined.

The stress-free management of recipients before and after embryo transfer is crucial to success. There appears to be too little emphasis on genotype of the recipient, as exemplified by wide variations in pregnancy rates and embryo survival among various strains of mice subjected to embryo transfer (Fekete 1947). Such diversity in success rate within a species emphasizes the importance of identifying a closely related species when considering interspecies embryo transfer. There is little doubt that less priority has been given to studying the failures of the recipient rather than the donor. Too often and for unknown reasons, the transfer of embryos, particularly in polytocous species, results in fewer offspring born than embryos transferred or no term pregnancy at all.

Technical Considerations

The need for close synchronization of estrus and ovulation between the donor and recipient is well established for most species. Optimally, the objective should be nearly exact synchrony, avoiding differences exceeding 24 hours. However, the tolerance limits appear dependent both on species as well as on the stage of embryo development.

One characteristic of certain carnivores (including the cat, mink, and ferret) is advantageous for embryo recovery and transfer and IVF. Because ovulation is nonspontaneous, that is, it occurs only after a copulatory or hormonal stimulus, the timing of oocyte and embryo recovery is controllable and even predictable. Artificial insemination or oocyte aspiration, for example, can be timed with injections of human chorionic gonadotropin (hCG) or gonadotropin-releasing hormone (GnRH), injectable compounds capable of inducing ovulation of mature follicles. Other exogenous gonadotropins (follicle stimulating hormone, FSH; pregnant mare serum gonadotropin, PMSG) can be used in the absence of natural estrus to stimulate estrous behavior and ovarian activity. These drugs usually induce development of a greater than normal number of ovarian follicles which can result in a superovulatory effect. Although in theory more ova and, thus, embryos are recovered, exogenous FSH and PMSG can produce cystic ovaries and prolonged intervals of estrous behavior. Repeated use of PMSG also has been associated with an immunological response, the injected animal developing antibodies capable of crossreacting with its own endogenous gonadotropins, thereby rendering the animal sterile. Generally, however, the repeated, judicious use of low dosages of FSH or even PMSG at prolonged intervals fails to induce a completely refractory response, at least in the cat. Of primary concern is the marked variability among species and among individuals within species in ovarian response to a given gonadotropin treatment. There is also increasing evidence that fertilization rates and embryo quality are compromised in females treated with these gonadotropins, in part due to abnormal endocrine status (Wildt et al. 1986), poor sperm transport (Wildt et al. 1987), and chromosomal aberrations in the oocyte (Fujimoto et al. 1974; Maudlin and Fraser 1977). The major limitation to improving efficiency in embryo recovery in women, domestic livestock, laboratory animals, and wildlife species is the lack of one or more consistently reliable, commercially available gonadotropin preparations.

Derivatives of the compound prostaglandin are valuable in synchronizing estrus and ovulation in various ungulates. Prostaglandin $F_{2\alpha}$ is luteolytic in these species, causing a rapid demise of the corpus luteum and, therefore, progesterone secretion. A decline in progesterone permits a resurgence in hypothalamo-pituitary release of hormones conducive to new follicle development and another period of estrus. To our knowledge, the influence of prostaglandins in the ferret has gone unstudied. However, the use of prostaglandin $F_{2\alpha}$ for the routine induction of luteolysis is ineffective for day 1 to day 30 canine corpora lutea, but is capable of inducing luteolysis and abortion from midgestation to term pregnancy (Concannon and Hansel 1977). Prostaglandin $F_{2\alpha}$ is ineffective in causing regression of the recently formed cat corpus luteum (Shille and Stabenfeldt 1979; Wildt et al. 1979).

Modern, well-equipped facilities are mandatory to a successful embryo program. Because living cells are being recovered and manipulated, it is crucial that at least a very clean, if not sterile, laboratory environment be maintained. The procedures of embryo recovery and transfer and IVF require general anesthesia of the donor and eventually the recipient. Although there is always risk in chemical immobilization, modern veterinary technology has advanced to allow repeated, safe anesthetic procedures of healthy females even over relatively brief intervals.

In large hoofstock species, embryo recovery is traditionally approached either surgically via a paralumbar incision or nonsurgically by inserting a flushing catheter through the cervix and into the uterus per rectal manipulation. Unfortunately, no nonsurgical embryo retrieval procedures have been reported for carnivores. Therefore, the total number of embryo recoveries is dictated by the number of major surgeries the female can withstand. Repeated manipulations of the reproductive tract, especially the oviducts, can easily result in tissue adhesions, severely compromising future embryo recovery attempts or the female's ability to conceive naturally. With extremely careful surgical approaches as many as 2 to 4 embryo recoveries could be accomplished before adhesions would be inhibitory.

Specific anatomical configurations also can influence access to the reproductive organs. The ovary of certain species, including the dog and ferret, is encapsulated in an opaque, peritoneal pouch, the ovarian bursa. The latter normally contains an opening too small for the ovary to be either extruded or easily observed by laparoscopy. Furthermore, the cranial aspect of the ferret's reproductive tract, including the ovarian bursa, is contiguous with a rather extensive fat pad, probably making direct manipulation of the tract an even more difficult procedure.

For IVF attempts in most species, mature oocytes are retrieved either by oviductal flushes or by follicle aspiration at laparotomy or preferably at laparoscopy. The latter technique has been used for this purpose in a number of laboratories (Wildt et al. 1986b). Laparoscopic aspiration involves identification of the mature ovarian follicle using a laparoscope inserted through a midline incision less than 1 cm long. Under direct laparoscopic observation and using a vacuum pump attached to a flask containing medium, a needle is used to puncture and aspirate oocytes from each mature follicle. Oocytes judged morphologically mature (germinal vesicle breakdown, polar body extrusion, or cumulus mass expansion) are fertilized with sperm collected by masturbation, an artificial vagina, or electroejaculation. Therefore, the primary advantage of the laparoscopy and IVF technique is that it is a less invasive approach for recovering the female's genetic material. Whether this technique can be applied to the ferret remains to be determined since the ovarian bursa precludes direct access to the ovarian surface. In the dog, this problem has been partially circumvented by minor laparoscopic surgery allowing exposure of the ovary (Wildt et al. 1977).

The disadvantage of IVF compared to embryo recovery and transfer is the degree of sophistication required for success. Both systems require specialized media, but for IVF the medium must sustain oocyte and spermatozoal viability as well as promote fertilization and cleavage. Whereas embryos recovered from a donor can be transferred immediately to the recipient, oocytes require detailed care and maintenance in a specialized artificial environment for at least 24 hours before transfer. The variation in embryo and oocyte sensitivity to various media and temperatures is significant and species dependent. For the ferret, these comparative norms have not been established. Furthermore, there is some published information that ferret ova readily undergo parthenogenetic cleavage, the formation and division of blastomeres in the absence of fertilization. Chang (1950, 1957) reported that 43 to 60% of noninseminated ferret ova cleaved spontaneously *in vivo*, most likely as a result of oocyte aging. Because of a haploid genotype, these "embryos" rarely develop beyond 4 to 8 cells and never result in offspring.

The assessment and handling of embryos and oocytes require sophisticated and expensive equipment including a stereomicroscope with high resolution optics allowing detailed morphological assessments; a tissue culture incubator permitting control of temperature and gas constituents; and an array of sterile cultureware, instruments, and finely drawn pipettes. If IVF is to be considered, additional equipment is needed for evaluating, counting, and washing spermatozoa. If atraumatic oocyte recovery is to be used, a major investment in laparoscopy equipment is required.

For either procedure the resulting embryos must be transferred safely to the suitably synchronized recipient. From a management perspective, the early luteal-phase (postovulatory) uterus of the recipient is particularly vulnerable to infection. Contaminated transfer medium, poor surgical technique, or extreme manipulation of the tract all are likely causes of pregnancy failure in recipients. The choice of transfer site, to oviduct or uterus, is largely governed by the stage of embryo development: 2- to 16-cell embryos are placed in the oviduct, while morulae and blastocysts are inserted into the uterine horns. In either case, it is important that corpora lutea be present on the ovary ipsilateral to the transfer. Because the transfer procedure requires only a single puncture of the reproductive tract, there is less concern of stimulating adhesion formation in the recipient. Alternatively, embryos could be transferred into the uterine horns of the recipient using a laparoscopic transabdominal approach, similar to that reported for sheep (Schiewe et al. 1984). Similarly, we have successfully cannulated the uterine horns of the dog and cat laparoscopically, directing a needle through the abdominal wall and into a predetermined site, a potential channel for embryo deposition (Wildt 1980). Because of the difficulty in transversing the cervix, no offspring have resulted in carnivores from nonsurgical embryo transfer to the recipient.

In polytocous animals, the number of embryos transferred can have a marked influence on prospects for survival (Adams 1982). In some species, such as the pig, a minimum of 4 embryos is necessary to maintain the early pregnancy (Day et al. 1967). In the rabbit, the pregnancy rate is greater in recipients receiving 5 embryos rather than 2 (Adams 1970). Although there is no evidence that time of day influences embryo transfer success (Adams 1982), it is likely that the level of stress experienced by the recipient is inversely proportional to pregnancy rate. Unfortunately, few data are available on the effects of various sedatives or anesthetics on the ability of the recipient to develop a pregnancy initiated by embryo transfer.

The greatest determinant of a successful embryo recovery and transfer or IVF program is the technical competence of the scientific team. These procedures, particularly oocyte and embryo handling and assessment, require manual dexterity and coordination that evolve only after months, or even years, of extensive experience.

Embryo Recovery and Transfer in Carnivores

Embryo recovery and transfer attempts have been reported only in several species of canids, felids, and mustelids. In most studies embryos were collected from or transferred to animals which had experienced a natural estrus. In one exception, Goodrowe et al. (1986) demonstrated that embryos recovered from an FSH-treated domestic cat and surgically transferred to the uterine horns of an FSH-treated recipient resulted in live-born offspring.

The first and only report of embryo transfer in the dog was published in 1978 using naturally estrous, artificially inseminated bitches (Kraemer et al. 1979). After inducing anesthesia, each uterine horn was flushed with Ham's F-10 medium containing 25 mM Hepes buffer, 10% heat-treated fetal calf serum plus penicillin and streptomycin. At laparotomy, 50 ml of medium was introduced into the uterine lumen at the tip of the horn using a plastic catheter and collected via a cannula inserted into the uterine body. Within 45 minutes the embryos in 1 to 5 μl of medium were transferred surgically to the uteri of recipients that had shown estrus within plus or minus 4 days of onset of estrus in the donors. A total of 72 embryos were recovered on 26 occasions. Thirty-seven embryos transferred to 7 recipients resulted in 3 pregnancies and the birth of 3 litters.

Embryo collection also has been accomplished in gonadotropin-treated bitches (Archbald et al. 1980). Females treated with PMSG (44 IU/kg body weight) for 9 consecutive days were allowed to breed naturally on the second day of estrus. Eight to 12 days later, each dog was laparotomized, and a phosphate-buffered saline solution was flushed through each horn into a catheter inserted into the cervix. Sixty-nine embryos (ranging from the 2-cell to morula stage) and 9 unfertilized ova were collected from 4 and 2 animals respectively. No data were provided concerning embryo culture or transfer.

Embryos also have been collected from one nondomestic canid, the silver fox (Pearson and Enders 1943). This study concentrated on the timing of ovulation and histological descriptions of embryo development. No embryo transfers were attempted.

The first account of successful embryo transfer in the cat was accomplished with naturally estrous donors and recipients (Kraemer et al. 1979). Donors were allowed to mate *ad libitum* for 1 or 2 days while recipients were induced to ovulate with 2 intramuscular injections of 250 IU of hCG or by mating with a vasectomized male. Surgical collections were performed via laparotomy 6, 7 or 9 days following the last breeding. The flush solution consisted of Ham's F-10 medium containing 25 mM Hepes buffer. Recovered embryos were maintained approximately 30 minutes at 32°C before surgical transfer to synchronized recipients (plus or minus 1 day). Forty-seven embryos were collected during 17 attempts and were transferred into 9 recipients. Four pregnancies resulted and 3 litters were live-born.

More recently, embryo collection and transfer have been accomplished in domestic cats mated for 3 days following induction of estrus with follicle stimulating hormone (2.0 mg/animal/day for 5 days) (Goodrowe et al. 1988a). Four to 7 days after the last breeding each uterine horn was flushed at laparotomy with 30 ml of Ham's F-10 plus 5% heat-treated fetal calf serum. A total of 43 and 82 embryos and unfertilized ova were collected from naturally and artificially induced estrous females respectively. Morphological assessments indicated that a greater proportion of poor- or degenerate-quality embryos and unfertilized ova were collected from the FSH-treated (51.6 and 24.4% respectively) compared to the naturally estrous (21.6 and 2.3% respectively) cats. The gonadotropin-treated females also produced fewer morula- and blastocyst-stage embryos than their naturally cycling counterparts. Embryos from FSH-treated cats, however, were capable of further development as evidenced by 3 pregnancies and the birth of 2 litters following 15 transfer attempts to hormonally synchronized recipients.

In a pilot study, Kraemer (1983) reported collecting 3 morulae during 5 attempts from a lioness in natural estrus. The embryos were frozen and thawed, but no pregnancy resulted after autotransfer back to the donor.

Artificial insemination and subsequent embryo collection in the common ferret were accomplished by Chang and Yanagimachi (1963). Epididymal spermatozoa, collected after castration, were placed in the ovarian capsule of naturally estrous females at various intervals before and after ovulation. A total of 278 embryos and unfertilized ova were collected from 47 dissected oviducts 1.5 to 78 hours after ovulation. To develop a fertilizing capacity, ferret spermatozoa were found to require 3.5 to 11.5 hours of residence time in the female reproductive tract. Therefore, the proportion of penetrated oocytes (54 to 62%) did not differ when insemination into the ovarian capsule occurred from 6 hours before to 12 hours after ovulation. When sperm were deposited

24 to 36 hours following ovulation, fertilization rates were decreased significantly (14 to 30%).

Embryo transfer in the ferret with subsequent fetal development was described by Chang (1968). Fifty-one morula- to blastocyst-stage embryos were transferred to the uterine horns of hCG-treated, naturally estrous females 6 to 8 days after mating. Implantation was observed in all 8 recipients, even in asynchronous transfers (day 8 embryos to a day 6 uterus, and day 6 embryos to a day 8 uterus). One female allowed to gestate to term delivered 4 live young.

Whittingham (1975) reported the culture of ferret embryos from the 1-cell to the blastocyst stage in a defined medium, but provided little other information.

Although Chang (1968) had no success in transferring 11 mink embryos to 2 mink recipients, Zheleznova and Golubitsa (1979) reported a 25% pregnancy rate after transferring blastocyst-stage embryos between the uteri of suitably synchronized mink.

Heterologous gamete interaction also has been studied in mustelids. Chang (1965, 1968) investigated the possibility of interspecific insemination and embryo transfer between the ferret and mink (a delayed implanter). Although mink spermatozoa were capable of fertilizing ferret ova with low rates of implantation, the reciprocal experiment failed. A similar interesting phenomenon occurred when interspecies embryo transfers were performed between the ferret and mink. Ferret embryos transferred to the mink uterus survived only 6 to 10 days, degenerating without implantation. Conversely, mink embryos were capable of undergoing implantation when transferred to the ferret uterus, but degenerated 4 to 8 days after penetration of the endometrium.

The development of ferret embryos in the rabbit oviduct and uterus also has been investigated (Chang 1966). A 2-day interval in the oviduct supported embryonic growth; however, after flushing on the 4th day, no embryos were recovered. The rabbit uterus proved to be an unfavorable culture environment. A subsequent study demonstrated that the culture of ferret embryos in the rabbit oviduct severely compromised later developmental capacity (Chang et al. 1971).

In Vitro Fertilization in Carnivores

Successful *in vitro* fertilization with subsequent birth of live offspring has been accomplished in the rabbit (Chang 1959), mouse (Whittingham 1968) rat (Toyoda and Chang 1974), human (Steptoe and Edwards 1978), cow (Brackett et al. 1982), pig (Cheng et al. 1986), sheep (Cheng et al. 1986), cat (Goodrowe et al. 1988b), and several species of non-human primates (Clayton and Kuehl 1984; Bavister et al. 1984; Bal-maceda et al. 1984). Despite continuous research efforts, only the most common laboratory animal or human systems seem to give repeatable pregnancy results. Certainly the greatest progress in IVF has occurred in the human infertility field, with this system now serving as a model for other species.

In the Carnivora, IVF has been accomplished only in the domestic dog and cat. Live-born offspring have been reported only from trans-ferring *in vitro* fertilized cat embryos (Goodrowe et al. 1988b).

The *in vitro* fertilization of dog oocytes matured *in vitro* has been reported by Mahi and Yanagimachi (1976, 1978). Although germinal vesicle breakdown was promoted by oocyte culture in BWW medium (Mahi and Yanagimachi 1976) the rate of sperm penetration of the zona pellucida or swelling of the sperm head within the vitellus was not influ-enced by the meiotic status of the oocyte. However, modification of the original salt, buffer, glucose, and bovine serum albumin components to give a defined canine capacitation medium resulted in a pronounced increase in the incidence of acrosome-reacted spermatozoa and sperm-penetrated oocytes (Mahi and Yanagimachi 1978).

In vitro fertilization of feline ova was first reported by Hamner et al. (1971). Following ovulation induction using PMSG (150 IU) and hCG (50 IU), oocytes were flushed from the oviducts and placed in culture in Brackett's medium in a 37°C, 5% CO_2 in air environment with *in vivo* capacitated spermatozoa ($15-75 \times 10^5$ spermatozoa/ml). A range of 33 to 90% of the ova were judged fertilized, and normal cleavage to the 16-cell stage occurred by the third day of culture.

Bowen (1977) collected oviductal oocytes from hCG-treated, natu-rally estrous cats. Mature ova were placed in culture (modified Ham's F-10 or BSW medium; 5% CO_2 in air at 38°C) with spermatozoa ($5.5-23 \times 10^5$ spermatozoa/ml) collected from the ductus deferens of castrated males. A comparison of the 2 different media indicated no differences in cleavage rates, with 70 to 100% of the ova undergoing division and some developing to the expanded blastocyst stage. Control (noninsemi-nated) oocytes failed to undergo parthenogenetic cleavage.

Niwa et al. (1985) recently reported *in vitro* fertilization of oocytes collected from domestic cats treated with PMSG (150 IU) and hCG (100 IU). The oocytes were flushed from the oviducts with a modified Kreb's-Ringer bicarbonate medium and placed in culture (5% CO_2 in air, 37°C culture environment) with $0.2-1.8 \times 10^6$ motile epididymal sper-matozoa per ml. Observations on the early precleavage events of fertil-ization demonstrated that spermatozoal capacitation required 15 to 20 minutes, while decondensation and swelling of the sperm head occurred as early as 30 minutes after insemination.

More recently in our laboratory, *in vitro* fertilization of feline follicular oocytes has resulted in morula-stage embryos (Goodrowe et al. 1988b). Females treated with PMSG (150 IU) and hCG (100 IU) 80 hours later were subjected to laparoscopy for follicular aspiration. The resulting mature and cumulus-oocyte complexes (mean = 11.5/female) were placed in a modified Kreb's-Ringer bicarbonate medium (Niwa et al. 1985) with freshly ejaculated-washed spermatozoa (2×10^5/ml) in a 5% CO_2 in air, humidified 37°C environment. Following a 20- to 24-hour culture period, cleavage rate averaged 45.2%. Five of 6 cats receiving 6 to 18 IVF embryos became pregnant and delivered from 1 to 4 kittens per litter. More recent data indicated that increasing the PMSG and hCG interval to 84 hours increased the rate of IVF to approximately 75% (Miller et al. 1988).

Possible Embryo Strategy for the Ferret

Our discussion has primarily emphasized the pitfalls and lack of baseline embryo data, not only for the ferret but for carnivores in general. There are, however, several possible scenarios which, if implemented in a sound and scientifically managed fashion, could suggest the eventual application of embryo technology to the black-footed ferret. One such approach is provided to serve as a potential guide for those responsible for species management.

The major prerequisite to embryo recovery and transfer or IVF in the black-footed ferret is the development of technical expertise in a suitable animal model counterpart, that is, a common species of ferret. The practicality and risks of certain manipulatory procedures (gonadotropin therapy, artificial insemination, surgical embryo recovery, and laparoscopic recovery of oocytes) can therefore be defined in the more common species first. Second, the model concept establishes reproductive norms for the species serving as the recipient for black-footed ferret embryos.

The responsibility of the ferret embryo program should be delegated to investigators with extensive embryo research experience. The principal investigators do not necessarily have to be familiar with a carnivore embryo system, although such expertise would be helpful. More important, these scientists should have documented credentials in the areas of hormonal therapy, gamete interaction, and the practical aspects of embryo recovery, assessment, culture, and transfer. A veterinary staff to

advise on anesthesia requirements and to participate in surgical procedures is essential. The animals must be maintained in facilities managed by wildlife professionals who can minimize environmental stress
factors detrimental to captive artificial breeding. A modern, clean laboratory (preferably in the building where the animals are maintained)
with technologically advanced equipment is mandatory.

A multidirectional approach using the common ferret is suggested to
promote optimal efficiency. For discussion purposes, 3 major research
areas are outlined, each component consisting of a number of individual
but complementary studies:

1. *Influence of gonadotropin therapies.* First, the normalcy of ovum or
embryo quality in naturally estrous females following ovulatory doses of
hCG or GnRH must be determined. One assessment option is to allow
hormonally treated animals to deliver offspring after mating. Alternatively, follicular oocytes or oviductal embryos could be collected from
hCG- or GnRH-treated ferrets and their viability analyzed after chromosomal analysis, IVF (in the case of oocytes), culture, or transfer to
recipients. Second, to overcome the problem of seasonal breeding, the
behavioral-ovarian responsiveness to exogenous gonadotropins requires study. Critical factors include: the type of gonadotropin (PMSG
versus FSH); number, dosage, and route of injections; and the normalcy
of behavioral estrus, ovarian morphology, and embryo quality in the
hormone-treated ferret. Because of the possible immunological risks of
repeated gonadotropin treatments, observations of ovarian responsiveness and embryo quality should be made when females are hormonally treated on 2 or more occasions. For maximal efficiency, initial studies should concentrate on ovarian-behavioral activity after various
hormonal treatments without flushing the reproductive tract. After a
hormonal treatment is proven, differential evaluations of embryo
viability can be implemented.

2. *Efficacy of embryo collection, culture, and transfer.* This area should
emphasize the atraumatic recovery of viable, early-stage embryos followed by transfer to suitably synchronized recipients. Principal issues to
be addressed include: the accessibility of the oviducts and uterine horns
for embryo-retrieval attempts; embryo-recovery efficiency after flushing the oviducts versus uterine horns; the safety and risk factors associated with surgical embryo recovery; the optimal medium for flushing
the reproductive tract and the short-term maintenance or transfer of
embryos; stage of embryo development optimal for achieving a pregnancy after embryo transfer; and the critical relationship of synchronization of the embryo donor and recipient as well as the potential
for a hormonally treated recipient to successfully serve as an embryo

surrogate. On a practical basis, studies would involve flushing the distinct segments of the reproductive tract (oviduct versus uterine horn) at different intervals after mating females in natural or hormonally induced estrus. Several of the flush and holding media previously used in other carnivore studies can be tested for maintaining short-term embryo viability before transfer. Selected females also can be designated for repeated embryo recoveries to assess the extent of adhesion formation after serial collections. Viable high-quality embryos can be transferred by laparotomy or laparoscopy to recipient females. The test group of recipients could be 1 of 2 types synchronized by: (1) coincident estrus with the donor and ovulation stimulated by hCG, GnRH, or copulation with a vasectomized male; or (2) FSH- or PMSG-induction of sexual behavior and follicle maturation followed by hCG-, GnRH-, or copulation-induced ovulation.

3. *Potential of IVF in combination with embryo transfer.* If a technique could be developed for modifying the ovarian bursa of the ferret, then the IVF approach could be attractive. In theory, the oocyte donor is treated with gonadotropin hormones and then subjected to follicular aspiration using laparoscopy. Spermatozoa collected from an electroejaculated male are washed and placed in various test media containing mature oocytes. Fertilization is judged microscopically and the embryo then transferred by laparotomy or laparoscopy to a recipient female. The first effort, therefore, must concern the feasibility of surgically exposing the ovary from the ovarian bursa. Such an approach may be complex or as simple as a laparoscopic lengthening of the natural bursal opening (Wildt et al. 1977). Subsequent major points of study then include: the interval between hormone treatment and ovum recovery since oocyte maturation is critical to IVF success; sperm capacitation requirements and numbers needed for IVF; assessing the actual fertilization event since the dark cytoplasm of the ferret oocyte will make pronuclear identification difficult; spontaneous cleavage of noninseminated oocytes (parthenogenesis) cultured *in vitro;* and determining the ability of embryos formed by IVF to result in live-born offspring.

Findings from these 3 areas will provide a data base for making logical decisions about the potential of embryo technology in the black-footed ferret. Again, this scenario is presented only as a general guide for one possible research strategy. Perhaps its major benefit is to reemphasize the complexity of the technology for preserving an endangered species. Other techniques—including the monitoring of reproductive activity and the collection, freezing, and artificial insemination of spermatozoa (which are discussed in other chapters)—are potentially critical to the success of an embryo program. Embryo cryopreservation is highly effective in certain species, but offers little immediate benefit in alleviating the current crises, particularly since there are no reports to date of frozen-thawed carnivore embryos resulting in live-born offspring.

Certainly not all concerns are biological. Whether embryo technology will have a favorable impact on the black-footed ferret can only be determined after financial resources become available. The illustrative strategy presented here poses numerous basic questions that are pre-emptively critical to practical success. Considerable financial resources are needed to attract professional expertise, to support sophisticated technical needs, and to supply sufficient numbers of model animals for making valid, scientific studies of the issues.

References

Adams, C. E. 1970. Maintenance of pregnancy relative to the presence of few embryos in the rabbit. *J. Endocr.* 48:243–50.

———. 1982. Egg transfer in the rabbit. In *Mammalian egg transfer*, ed. C. E. Adams, 30–48. Boca Raton: CRC Press.

Archbald, L. F., et al. 1980. A surgical method for collecting canine embryos after induction of estrus and ovulation with exogenous gonadotropins. *Veterinary Medicine/Small Animal Clinician* Feb.:228–38.

Balmaceda, J. P., et al. 1984. Successful *in vitro* fertilization and embryo transfer in cynomologous monkeys. *Fert. Ster.* 42:791–95.

Bavister, B. D., et al. 1984. Birth of rhesus monkey infant after *in vitro* fertilization and nonsurgical embryo transfer. *Proc. Nat. Acad. Sci.* 81:2218–22.

Blandau, R. J. 1980. *In vitro* fertilization and embryo transfer. *Fert. Ster.* 33:3–11.

Bowen, R. A. 1977. Fertilization *in vitro* of feline ova by spermatozoa from the ductus deferens. *Biol. Reprod.* 17:144–47.

Brackett, B. G., et al. 1982. Normal development following *in vitro* fertilization in the cow. *Biol. Reprod.* 27:147–58.

Chang, M. C. 1950. Cleavage of unfertilized ova in immature ferrets. *Anatom. Rec.* 108:31–44.

———. 1957. Natural occurrence and artificial induction of parthenogenetic cleavage of ferret ova. *Anatom. Rec.* 128:187–200.

———. 1959. Fertilization of rabbit ova *in vitro*. *Nature* 184:466–67.

———. 1965. Implantation of ferret ova fertilized by mink sperm. *J. Exp. Zool.* 160:67–80.

———. 1966. Reciprocal transplantation of eggs between rabbit and ferret. *J. Exp. Zool.* 161:297–305.

———. 1968. Reciprocal insemination and egg transfer between ferrets and mink. *J. Exp. Zool.* 168:49–60.

Chang, M. C., and R. Yanagimachi. 1963. Fertilization of ferret ova by deposition of epididymal sperm into the ovarian capsule with special reference to the fertilizable life of ova and capacitation of sperm. *J. Exp. Zool.* 154:175–88.

Chang, M. C., et al. 1971. Development of ferret eggs after 2 to 3 days in the rabbit oviduct. *J. Reprod. Fert.* 25:129–31.

Cheng, W. T. K., et al. 1986. *In vitro* fertilization of pig and sheep oocytes maintained *in vivo* and *in vitro*. *Theriogen.* 25:146.

Clayton, O., and T. J. Kuehl. 1984. The first successful *in vitro* fertilization and embryo transfer in a nonhuman primate. *Theriogen.* 21:228.

Concannon, P. W., and W. Hansel. 1977. Prostaglandin $F_{2\alpha}$ induced-luteolysis, hypothermia and abortions in Beagle bitches. *Prostaglandins* 13:533–38.

Day, B. M., et al. 1967. Embryo numbers and luteal maintenance during early pregnancy in swine. *J. Anim. Sci.* 28:1499–1504.

Fekete, E. 1947. Differences in the effect of uterine environment in the DBA and C57 black strains of mice. *Anatom. Rec.* 98:409–16.

Fujimoto, S., et al. 1974. Chromosome abnormalities in rabbit preimplantation blastocysts induced by superovulation. *J. Reprod. Fert.* 40:177–81.

Goodrowe, K. L., et al. 1988a. Comparison of embryo recovery, embryo quality, oestradiol-17β and progesterone profiles in domestic cats at natural and induced-oestrus. *J. Reprod. Fert.* 82:553–61.

Goodrowe, K. G., et al. 1988b. Developmental competence of domestic cat follicular oocytes after fertilization *in vitro*. *Biol. Reprod.* 39:355–72.

Gwatkin, R. B. L. 1977. *Fertilization mechanisms in man and mammals*. New York: Plenum Press.

Hamner, C. E., et al. 1971. Cat (*Felis catus*) spermatozoa require capacitation. *J. Reprod. Fert.* 23:477–80.

Kraemer, D. C. 1983. Intra- and interspecific embryo transfer. *J. Exp. Zool.* 228:363–71.

Kraemer, D. C., et al. 1979. Embryo transfer in the nonhuman primate, feline and canine. *Theriogen.* 11:51–62.

Mahi, C. A., and R. Yanagimachi. 1976. Maturation and sperm penetration of canine ovarian oocytes *in vitro*. *J. Exp. Zool.* 196:189–96.

———. 1978. Capacitation, acrosome reaction, and egg penetration by canine spermatozoa in a simple defined medium. *Gamete Research* 1:101–9.

Maudlin, I., and L. R. Fraser. 1977. The effect of PMSG dose on the incidence of chromosomal anomalies in mouse embryos fertilized *in vitro*. *J. Reprod. Fert.* 50:272–80.

Miller, A. M., et al. 1988. Influence of gonadotropin source and treatment interval on *in vitro* fertilization of cat follicular oocytes. *Biol. Reprod.*, Suppl. 38:73.

Newcomb, R. 1980. Investigation of factors affecting superovulation and non-surgical embryo recovery from lactating British Freisian cows. *Vet. Rec.* 106:48–50.

Niwa, K., et al. 1985. Early events of *in vitro* fertilization of cat eggs by epididymal spermatozoa. *J. Reprod. Fert.* 74:657–60.

Pearson, O. P., and R. K. Enders. 1943. Ovulation, maturation and fertilization in the fox. *Anatom. Rec.* 83:69–83.

Polge, C., and S. M. Willadsen. 1978. Freezing eggs and embryos of farm animals. *Cryobiol.* 15:370–73.

Rogers, J. B. 1978. Mammalian sperm capacitation and fertilization *in vitro:* A critique of methodology. *Gamete Research* 1:165–223.

Schiewe, M. C., et al. 1984. Laparoscopic embryo transfer in domestic sheep: A preliminary study. *Theriogen.* 22:675–82.

Schmidt, P. M., et al. 1985. Viability of frozen-thawed mouse embryos is affected by genotype. *Biol. Reprod.* 32:507–14.

Seidel, G. E. 1981. Embryo transfer procedures with cattle. In *Fertilization and embryonic developments in vitro,* ed. L. Mastroianni and J. D. Biggers, 324–54. New York: Plenum Press.

Shille, V. M., and G. H. Stabenfeldt. 1979. Luteal function in the domestic cat during pseudopregnancy and after treatment with prostaglandin. *Biol. Reprod.* 21:1217–23.

Steptoe, P. C., and R. G. Edwards. 1978. Birth after re-implantation of a human embryo. *Lancet* 2:366.

Toyoda, Y., and M. C. Chang. 1974. Fertilization of rat eggs *in vitro* by epididymal spermatozoa and the development of eggs following transfer. *J. Reprod. Fert.* 36:9–22.

Whittingham, D. G. 1968. Fertilization of mouse eggs *in vitro. Nature* 220:592–93.

———. 1975. Fertilization, early development and storage of mammalian ova *in vitro.* In *The early development of mammals,* ed. M. Balls and A. E. Wild, 1–24. New York: Cambridge Univ. Press.

———. 1979. *In vitro* fertilization, embryo transfer and storage. *British Med. Bull.* 35:105–11.

Wildt, D. E. 1980. Laparoscopy in the dog and cat. In *Animal laparoscopy,* ed. R. M. Harrison and D. E. Wildt, 31–72. Baltimore: Williams and Wilkins.

Wildt, D. E., et al. 1977. Laparoscopic exposure and sequential observation of the ovary of the cycling bitch. *Anatom. Rec.* 3:443–49.

Wildt, D. E., et al. 1979. Effect of prostaglandin $F_{2\alpha}$ on endocrine-ovarian function in the domestic cat. *Prostaglandins* 18:883–92.

Wildt, D. E., et al. 1986. Developing animal model systems for embryo technologies in rare and endangered wildlife. *Theriogen.* 25:33–52.

Wildt, D. E., et al. 1987. Seminal-endocrine characteristics of the tiger and the potential for artificial breeding. In *Tigers of the world: The biology, biopolitics, management and conservation of an endangered species,* ed. R. Tilson and U. S. Seal. Park Ridge, N.J.: Noyes.

Wright, R. W., and K. R. Bondioli. 1981. Aspects of *in vitro* fertilization and embryo culture in domestic animals. *J. Anim. Sci.* 53:702–29.

Zheleznova, A. I., and A. N. Golubitsa. 1979. Transplantation of blastocysts in the mink. *Anim. Breed. Abst.* 47:545 (#4970).

12

ROBERT W. ATHERTON, MONTE STRALEY, PATRICK CURRY, RICHARD
SLAUGHTER, WARREN BURGESS, AND ROBERT M. KITCHIN

Electroejaculation and Cryopreservation of Domestic Ferret Sperm

Application of technological advances in reproductive biology offer wildlife biologists new approaches to managing threatened or endangered species. Electroejaculation, cryopreservation of sperm, artificial insemination, and, to a degree, embryo transfer are widely practiced as effective means of improving livestock. Artificial insemination with cryopreserved semen is also used in treating infertility. These techniques are applied, with some success, in both zoos and various captive-breeding programs for endangered species (Seager 1983; Seal and Foose 1983). This chapter describes for the first time the cryopreservation of ferret sperm and recovery of post-thaw motility. These studies are still in progress, and we hope to develop an effective procedure for electroejaculation and cryopreservation of ferret sperm that can be used in the black-footed ferret captive-breeding program.

The sexually mature, domestic male ferrets used in this study were obtained from Marshall Farms (North Rose, N.Y.) and individually caged in animal facilities of the Department of Zoology and Physiology, University of Wyoming, Laramie. Animals were maintained on a controlled light cycle (with a mean of 16.8 foot candles for 13.5 hours per day), fed commercial dry cat chow twice daily, and provided water *ad libitum*. Experiments reported here were conducted from April 8 through July 8, 1986.

Electroejaculation

Ferrets were anesthetized with 60 mg Ketamine hydrochloride (Vetalar, Parke-Davis) before electroejaculation (Moreland and Glaser 1985). Length and width of the left and right testis were measured with vernier calipers before electroejaculation. Semen was collected in prewarmed 1.5-ml microcentrifuge tubes by electroejaculation according to Shump et al. (1976a, 1976b), with the following modifications. A longitudinal bipolar rectal electrode (9.1 mm in diameter) was inserted approximately 35 mm into the rectum. Electrical stimulations were applied beginning at 7 mA and 6.3 mV and in gradually increasing steps up to 41 mA and 12.6 mV as necessary. Stimulations of 2-seconds duration were applied through a rheostat with approximately 1-second pauses between successive stimulations. Five stimulations were made at the 1st through the 4th setting on the electroejaculator (Nicholson, Denver, Colo.), and if no semen was collected, up to 100 stimulations were made at the 5th setting. If no semen was obtained, 50 stimulations were given at the 6th setting and the current was then reduced to the 5th setting on the electroejaculator for additional stimulations. The rectal electrode was washed with 95% ethanol between use on animals. The volume of a ferret ejaculate is approximately 5 to 20 μl (Shump et al. 1976a). One hundred μl of extender was added to the ejaculate.

Cryopreservation Procedures

Semen was maintained in an incubator at 37°C for a maximum of 90 minutes until all samples were collected. Five μl of each sample were used for video analyses, and the remaining volume was placed in a water bath and cooled to 4°C in a low temperature incubator (Precision model 815, Chicago, Ill.). At 4°C, 25 μl of precooled extender containing 2 times the desired glycerol concentration was added by drops to the extended sample every 15 minutes for 1 hour until the final concentration of glycerol was attained. A 25-μl solution of 10 mM ATP (adenosine triphosphate) per extender was added once to 100 μl of extended sample, yielding a 2-mM ATP concentration. The extended samples were pipetted into precooled 250-μl cryopreservation straws. The samples were then equilibrated at 4°C for 1 hour and placed in nitrogen vapors approximately 7 cm above the level of liquid nitrogen (approximately −60°C). Between 1 and 2 hours later, they were immersed and stored in liquid nitrogen. The rate of cooling the samples from 37°C down to 4°C varied with experiments.

To obtain a particular cooling rate the microcentrifuge tubes were placed in a plastic test tube rack, mounted in a 1-liter water cooling chamber at 37°C and cooled to 4°C. To obtain a faster or slower cooling rate, the low temperature incubator was programmed accordingly. In the last set of experiments, extended samples were cooled at a maximum of 10°C per hour in a programmable slow freeze apparatus (Canalco) from 4°C to −60°C, placed over liquid nitrogen vapors 1 hour, and then stored in liquid nitrogen a minimum of 48 hours before thawing.

Semen samples were thawed at room temperature, then removed from the cryopreservation straw and placed in a 37°C incubator for 10 minutes before a 5-μl sample was removed and examined by phase contrast microscopy for sperm concentration and motility.

Cryopreservation extenders used in these experiments included egg-yolk citrate (EYC), egg-yolk test (EYT), TC199-egg yolk, and commercial (glucose-nonfat dry milk) horse semen extender (E-Z Mixin 2-way equine semen extender: Animal Reproduction Systems, Chino, Calif.). Egg-yolk citrate extender contained 65 mM of citric acid and 60 mM of fructose (pH 7.4). The egg-yolk test extender consisted of 210 mM of TES (N-tris [hydroxymethyl] [methyl-2-aminoethane sulfonic acid]), 96 mM of TRIS (tris [hydroxymethyl] amino methane]) and 12 mM of fructose (pH 7.4). Egg-yolk tris contained 96 mM of TRIS, 65 mM of citric acid, and 60 mM of fructose (pH 7.4). Egg yolk (20% v/v) was added to each buffer. In addition, tissue culture medium TC199 with 10% fetal calf serum (GIBCO) with and without 20% fresh egg yolk was tested as a cryopreservative. Egg yolk (EY) was centrifuged at either 15,000 × g for 15 minutes or 2,000 × g for 5 minutes to remove suspended particulates which interfered with microscopic evaluation of sperm motility.

A 5-μl sample was placed in a Makler counting chamber maintained at 37°C and examined by phase microscopy at 160× to determine sperm per ml and motility (Makler 1978). Sperm were videotaped with a JVC GX-S700U color video camera and Panasonic AG-6010 videocassette recorder. Approximately 5- to 10-second recordings of 20 to 30 fields were made.

Whenever possible, a total of 200 sperm were scored for motility from videotapes on a Sony PVM-1261Q monitor at 420×. Sperm were classified as progressively motile if they showed forward motility, regardless of speed, or as circularly motile if they moved primarily in circles. Sperm which exhibited flagellar motion but were otherwise nonmotile were classified as wigglers. Nonmotile sperm neither moved nor showed flagellar movement. In this study the percentage of motile sperm includes progressively motile, circularly motile, and wiggling sperm. A series of similar experiments with slightly different protocols were carried out at Colorado State University by Bowen and co-workers (see Appendix).

Semen samples containing motile sperm were successfully collected in 93% (87/93) of all attempted electroejaculations. Electroejaculations were considered unsuccessful if the ejaculate contained no sperm or urine was present, which inhibited sperm motility. Semen was obtained after as few as 26 to as many as 308 electrical stimulations. A mean of 98.8 ± 8.4 (±SE) stimulations were required per semen sample collected. In general, no ill effects resulted from electroejaculation. One animal developed mild rectal bleeding, but after 2 weeks of antibiotic treatment, the animal and his semen were again normal.

Testis size, sperm concentration, and sperm motility varied between both individuals (Table 12.1) and collection dates (Table 12.2). Sperm concentration in ejaculates varied from 4.7 to 41.1 × 10^6 sperm per ml. Animals 1, 2, 8, and 10 produced ejaculates with relatively few sperm (Table 12.1). Animal 1 was sacrificed for biochemical and genetic analyses after 2 semen samples had been collected. Mean testis width and length were 17.1 ± 0.7 and 17.4 ± 0.8 mm respectively on the first collection and 12.2 ± 0.9 and 16.3 ± 1.0 mm on the final collection 91 days later (Table 12.2). There was no correlation between testis size and either sperm concentration or motility during this study.

The percentage of motility of fresh and post-thawed sperm is summarized in Tables 12.3 through 12.5. The overall percentage of motile sperm in fresh ejaculates ranged from 46 to 86% (Table 12.1), while post-thaw motility ranged from 0 to 48% (Tables 12.3–12.5).

Absence of EY from TC199 decreased sperm motility in fresh semen, and no post-thaw motility was found in these samples. When sperm were

Table 12.1. Individual Testis Size, Sperm Concentration, and Sperm Motility of Domestic Ferrets from April 8 to July 8, 1986

Individual	Samples (no.)	Average Testis Size		Million sperm/ml	Motile sperm (%)
		Width (mm)	Length (mm)		
1	2	13.8 ± 1.0[a]	21.1 ± 0.9	9.0	80.0 ± 6.0
2	8	14.7 ± 0.5	19.0 ± 0.8	4.7 ± 2.2	69.6 ± 7.4
3	8	15.4 ± 0.6	20.6 ± 1.0	14.7 ± 5.8	71.2 ± 10.4
4	7	14.6 ± 0.4	19.0 ± 0.5	19.1 ± 4.8	59.0 ± 9.3
5	7	15.5 ± 0.4	19.5 ± 0.5	13.8 ± 5.5	62.5 ± 13.8
6	7	15.0 ± 0.6	19.4 ± 0.9	21.8 ± 6.0	73.7 ± 8.8
7	9	14.3 ± 0.8	17.3 ± 0.8	26.8 ± 7.7	86.4 ± 5.0
8	10	13.8 ± 0.5	19.8 ± 1.1	11.3 ± 4.0	70.1 ± 8.5
9	8	14.1 ± 0.5	19.0 ± 0.8	41.1 ± 11.4	78.9 ± 6.5
10	9	12.3 ± 0.9	16.8 ± 1.2	5.5 ± 1.3	54.6 ± 17.1
11	10	14.8 ± 0.9	18.7 ± 0.9	26.6 ± 8.1	84.9 ± 2.6
12	10	16.4 ± 0.6	19.9 ± 0.8	24.6 ± 6.8	80.3 ± 6.1

[a]Mean ± standard error.

Table 12.2. Average Testis Size, Sperm Concentration, and Sperm Motility of Domestic Ferrets on Each Day of Collection from April 8 to July 8, 1986

Day	Samples (no.)	Average Testis Size		Million sperm/ml	Motile sperm (%)
		Width (mm)	Length (mm)		
1	7	17.1 ± 0.7[a]	17.4 ± 0.8	nd[b]	48.7 ± 9.9
9	3	nd	nd	8.0 ± 2.1	75.8 ± 9.5
16	5	14.6 ± 0.6	21.3 ± 0.4	30.0 ± 13.9	82.0 ± 11.4
23	12	15.0 ± 0.2	20.9 ± 0.8	26.4 ± 8.9	68.4 ± 12.3
30	12	15.0 ± 0.5	20.1 ± 0.6	14.4 ± 4.0	75.3 ± 8.0
37	6	16.1 ± 0.4	20.0 ± 0.5	4.3 ± 2.7	59.7 ± 13.1
42	5	15.5 ± 1.5	20.8 ± 1.3	17.0 ± 3.0	74.2 ± 17.5
49	6	14.0 ± 0.5	17.4 ± 0.9	11.2 ± 4.8	96.8 ± 1.1
52	5	14.5 ± 0.6	20.2 ± 1.0	17.2 ± 7.0	79.4 ± 5.9
56	6	15.6 ± 0.5	19.9 ± 0.7	25.0 ± 11.1	57.0 ± 14.0
66	5	13.2 ± 0.5	17.1 ± 1.0	30.8 ± 13.1	83.0 ± 3.3
70	5	14.2 ± 0.3	18.0 ± 0.5	26.7 ± 12.5	71.8 ± 6.1
73	5	13.6 ± 1.0	17.8 ± 0.8	28.6 ± 18.5	90.8 ± 3.4
79	5	13.8 ± 0.4	18.6 ± 0.7	34.0 ± 9.9	84.5 ± 3.1
91	11	12.2 ± 0.9	16.3 ± 1.0	21.1 ± 3.1	76.4 ± 3.8

[a]Mean ± standard error.
[b]nd = not determined.

extended in TC199 plus EY, 36% post-thaw motility was observed. This value is an estimate, however, since the absolute number of nonmotile sperm were obscured by egg-yolk particles. Neither low nor high speed centrifugation of the TC199 plus EY extender allowed post-thaw motility analysis (Table 12.3).

Motile sperm were also obtained following cryopreservation in EYC or EYT extenders. Up to 13% of the original motility was regained after cryopreservation in EYC or EYT, with higher motility observed for sperm frozen in EYT. When sperm were quickly frozen in EYC as pellets on dry ice, however, motility was less than in either EYC or EYT in straws (Table 12.4).

Using video analysis, the highest percentage of post-thaw motility was observed in commercial horse extender (Table 12.5). In horse extender, sperm often exhibited circular motility and sperm heads had a tendency to clump together before freezing. But circular motility and head agglutination largely disappeared after thawing. Approximately 19% of sperm frozen in horse extender were motile upon thawing, and semen samples frozen with 2 mM ATP exhibited almost twice (35.6%) these values upon thawing (Table 12.5).

Table 12.3. Sperm Motility in Fresh and Cryopreserved Domestic Ferret Semen in Tissue Culture Medium TC199 with and without 20% Egg-Yolk (EY) Extender

Extender	Cooling Rate[a] (°C/hr)	Glycerol (%)	Sperm/ml	Samples (no.)	Motile Sperm	
					Fresh (%)	Thawed (%)
TC199 without EY	11	3.8	6.3 ± 5.3[b]	3	45.7 ± 24.4	0
TC199 with EY						
Uncentrifuged	5	0	6.5 ± 0.5	2	95.5 ± 1.5	4.0 ± 4.0
	5	4.5	14.3 ± 9.9	3	98.0 ± 2.0	35.7 ± 18.0
Centrifuged	5	0	27.3 ± 11.3	6	85.7 ± 2.3	3.0 ± 1.0
(low speed)	9	0	37.3 ± 31.0	3	95.7 ± 2.8	30.3 ± 15.0
	5	3.0	43.3 ± 18.5	3	87.0 ± 3.6	1.0 ± 1.0
	5	3.0	7.0 ± 0.0	1	99.0 ± 0.0	6.0 ± 0.0
Centrifuged	10	0	29.3 ± 8.4	7	87.4 ± 3.2	2.9 ± 1.1
(high speed)	10	3.6	20.8 ± 12.3	4	81.5 ± 3.2	14.5 ± 8.2

[a]Rate of cooling from 37°C to 4°C.
[b]Mean ± standard error.

Table 12.4. Sperm Motility in Fresh and Cryopreserved Domestic Ferret Semen in Egg-Yolk Test (EYT) and Egg-Yolk Citrate (EYC) Extenders

Extender	Cooling Rate[a] (°C/hr)	Glycerol (%)	Sperm/ml	Samples (no.)	Motile Sperm	
					Fresh (%)	Thawed (%)
EYT	7	4.5	43.5 ± 15.4[b]	4	89.5 ± 4.1	12.8 ± 7.8
	6	6.0	14.4 ± 4.0	10	75.3 ± 4.1	5.8 ± 2.0
EYC	4	1.0	43.0 ± 32.0	2	83.5 ± 11.5	0
	4	3.0	19.0 ± 4.7	3	56.0 ± 16.8	0
	6	4.3	30.8 ± 9.1	6	79.8 ± 5.6	2.7 ± 1.0
EYC[c]		0	nd[d]	4	59.3 ± 18.7	4.3 ± 2.2

[a]Rate of cooling from 37°C to 4°C.
[b]Mean ± standard error.
[c]Samples fixed as a pellet by dropping onto dry ice after sample reached 4°C.
[d]nd = not determined.

Table 12.5. Sperm Motility in Fresh and Cryopreserved Domestic Ferret Semen in Commercial Horse Extender with and without 2 mM ATP

Horse Extender	Cooling Rate[a] (°C/hr)	Glycerol (%)	Sperm/ml	Samples (no.)	Motile Sperm Fresh (%)	Motile Sperm Thawed (%)
Without ATP	4	0	18.3 ± 10.9[b]	3	79.7 ± 10.3	27.0 ± 11.7
	5	0	18.0 ± 10.0	2	73.0 ± 13.0	9.5 ± 3.5
	7	0	4.6 ± 6.7	8	73.3 ± 4.1	18.8 ± 2.6
	4	3.0	18.5 ± 8.7	4	83.0 ± 3.3	21.8 ± 7.0
	5	3.0	7.5 ± 0.5	2	58.0 ± 13.0	15.0 ± 1.0
	5	4.5	15.0 ± 5.0	2	86.5 ± 3.5	21.0 ± 12.0
+2 mM ATP	5[c]	0	17.0 ± 2.0	2	78.5 ± 16.5	14.0 ± 8.0
	5[c]	3.0	32.0 ± 4.0	2	81.0 ± 3.0	45.0 ± 2.0
	5[c]	4.5	21.5 ± 1.5	2	81.5 ± 4.5	48.0 ± 4.0

[a]Rate of cooling from 37°C to 4°C.
[b]Mean ± standard error.
[c]Rate of cooling from 4°C to 0°C was 4°C/hr; from 0°C to −60°C was 10°C/hr.

Effects of glycerol on post-thaw sperm motility are also presented in Tables 12.3 through 12.5. In general, the highest percentage of sperm motility was found in semen frozen in extenders containing 3 to 4.5% glycerol. Samples cryopreserved in uncentrifuged TC199 plus EY, egg-yolk citrate, egg-yolk test, and horse extender were of particular interest, since the percentage of motile sperm usually increased with glycerol content within the range (1 to 6%) tested. Although horse extender (±2 mM ATP) gave the highest motility of extenders without glycerol, it is important to stress that these samples were cooled slowly during the liquid-solid phase transition between 4°C and −20°C as described below.

Finally, the rate of cooling samples from 37°C to 4°C was studied. The rate of cooling ranged from 4° to 11°C per hour. The highest percentage of motile sperm was in samples cooled at 4° to 7°C per hour (Tables 12.3–12.5). The highest percentage of post-thaw motility occurred in the last set of experiments in samples cooled at a maximum of 10°C per hour from 4°C to −60°C before being placed in liquid nitrogen. During the freezing transition (4°C to −5°C), the temperature was decreased from 4°C to 0°C over a 1-hour period (Table 12.5). In contrast, all other semen samples in earlier experiments were placed directly into liquid nitrogen vapors after they had equilibrated at 4°C and glycerol had been added.

This is the first report of methods for the cryopreservation of domestic ferret sperm. In this study, semen was collected once or twice weekly over a 3-month period. The volume of semen collected was estimated to be 5 to 20 µl and appears similar to volume previously reported. Sperm concentration per ml was also similar to that previously reported (Shump et al. 1976b).

Testis size, sperm concentration, and the percentage of motile sperm varied between both individuals and collection dates. While mean testis size was similar to that reported previously (Neal et al. 1977), in the current study males with small testes (less than 10 mm in width and length) produced little or no semen, but whether this can be extrapolated to immature males is unknown. Absence of a correlation between testis size and these parameters is not entirely surprising, since this study was performed during a 3-month period when the ferret testes were at their maximum size (Neal et al. 1977). Recent studies from our lab compare in detail the motility (Kitchin et al. 1988) and morphology (Curry et al. 1989) of fresh sperm from domestic and black-footed ferrets.

Our major objective was to develop a method for the cryopreservation of black-footed ferret sperm. During cryopreservation, sperm may be damaged in cooling (from 37°C to 4°C), freezing (from 4°C to −25°C), or thawing, with disruption of structural integrity of sperm membranes (Watson 1981) and loss of sperm motility (Lindemann et al. 1981). Poorly cryopreserved semen with a high proportion of structurally damaged or nonmotile sperm results in low fertilization rates during artificial insemination (Pilikian et al. 1982).

Cryopreservative agents contain a variety of substances, including egg yolk, milk or milk proteins, and glycerol, which provide protection to spermatozoa undergoing cold shock (membrane disruption) and exposure to low temperatures. Glycerol appears to stabilize sperm membranes and membrane-bound proteins within an appropriate range of concentrations (Douzou 1977), while higher glycerol concentrations disrupt sperm membranes (Drevius 1972). However, optimum glycerol content can vary considerably between species; it is about 3% for fox semen (Fougner et al. 1973) and about 12 to 16% glycerol for human spermatozoa (Pilikian et al. 1982).

Rate of cooling and freezing appears to be particularly important in maximizing post-thaw sperm motility. Samples frozen very slowly, particularly during the liquid-solid transition phase between 4°C and −20°C, gave higher post-thaw motility and progressive motility, even when the extender lacked glycerol. Bull, stallion, and ram spermatozoa are highly susceptible to cold shock, but when frozen slowly (less than 10°C/hr) motile sperm are obtained upon thawing (Blackshaw and Salisbury 1957).

The highest percentage of ferret sperm motility was obtained in semen frozen in commercial horse extender using straws. Since most samples cryopreserved in horse extender were cooled and frozen at a slow rate below 4°C, it is as yet unknown if other extenders may yield similar or better results if frozen by the same method. Data from cryopreservation of feline sperm (Dresser et al. 1987) coupled with successful artificial insemination suggest that the dry-ice-pellet procedure may be an important method to pursue (see Appendix).

The addition of ATP as an energy source significantly increases sperm motility of membrane-damaged bull sperm after cryopreservation (Lindemann et al. 1981). Addition of ATP to ferret sperm also increased post-thaw motility, but whether changes in sperm motility after the addition of ATP reflect membrane damage or some other factor has yet to be determined. Preliminary analysis from slow-motion video playback indicates that with 2 mM ATP, 4.5% glycerol, and slow cooling rates, up to 20% of all motile sperm after thawing (Table 12.5) were progressively motile.

TC199 medium has been used for the artificial insemination of ferrets

(Chang and Yanagimachi 1963). Therefore, this extender has been studied alone and with the addition of egg yolk and glycerol. Unfortunately, the extender aggregated, possibly due to the complex mixture of fetal calf serum, egg yolk, and glycerol, precluding it from accurate microscopic analysis.

P. BUDWORTH, R. P. AMANN, R. A. BOWEN

Appendix: Cryopreservation of Semen from Domestic Ferrets

Semen was collected by electroejaculation from 5 domestic ferrets, using techniques similar to those described above. Following collection, 100 μl of glycerol-free extender was added to the semen and mixed. Approximately 5 μl was then examined to estimate initial motility. In all experiments, the remainder of the extended semen was split into two 50-μl aliquots to allow use of split-ejaculate comparisons of glycerol concentrations. Seminal volume was too small to allow comparison of different extenders by the split ejaculate technique. Rather, semen from each male was rotated between extenders as part of a weekly collection schedule. Following initial extension, tubes containing semen were suspended in scintillation vials containing water at 35°C to provide a water jacket for cooling; this apparatus was then placed in a 500-ml beaker surrounded by crushed ice, a system that allowed cooling to 5°C in 2 hours. After all semen samples had been cooled to 5°C, they were glycerolated by addition of 3 aliquots (15, 15, and 20 μl) of glycerol-containing extender (2 times the desired final concentration) at 10-minute intervals. The samples were then equilibrated at 5°C for an additional 4 hours, frozen as pellets on dry ice, and stored in liquid nitrogen. Pellets were thawed by agitation in 50 μl of prewarmed, homologous extender, and progressive motility was estimated by a single experienced observer; samples with relatively high post-thaw motility were photographed using darkfield optics for photomotility analysis.

An initial experiment comparing egg yolk (EY)-lactose and EY-tris extenders, each with final glycerol concentrations of 3% or 6%, indicated that there were no dramatic differences between glycerol levels or between the 2 extenders. A second experiment was conducted (Table A.1) to compare EY-TRIS and EY-TEST (TRIS, TES and fructose buffer with 29% egg yolk), with 2% versus 4% glycerol (final concentration). There was no difference in post-thaw survival between 2% and 4% glycerol in either extender, but semen frozen in EY-TEST extender had significantly higher ($p < 0.01$) post-thaw survival than that frozen in EY-tris. This difference was maintained even after exclusion of data from 1 male.

Currently, we are comparing semen extended in EY-TEST with 2% glycerol and frozen in either 0.25 ml ministraws or as pellets; initial results suggest that these 2 packaging systems do not result in significantly different post-thaw survival.

Table A.1.   Visual Estimate of Percentage of Progressive Post-Thaw Motility of Ferret Semen Frozen in Pellets

	Egg yolk-Tris extender		Egg yolk-TEST extender	
Male	2% Glycerol	4% Glycerol	2% Glycerol	4% Glycerol
1	15	15	35	35
	10	15	nd[a]	nd
2	20	10	15	15
	<5[b]	10	30	30
4	<5	<5	35	35
	<5	<5	25	30
5	15	15	30	25
	20	10	20	10
6	15	15	nd	nd
	10	15	20	25
Mean	12	12	26	26

[a]nd = concentration of ejaculate too low to freeze.
[b]Scored as 5% for statistical analysis.

Two estrous ferrets were inseminated (by surgical placement into the uterus under ketamine anesthesia) with frozen-thawed spermatozoa from 1 ejaculate of 1 male. This semen was frozen in EY-TEST extender; 1 female received semen frozen with 2% glycerol and the other, semen frozen with 4% glycerol. Approximately 2,000,000 motile spermatozoa (35% progressive motility, the equivalent of 1 pellet) were inoculated into the body of the uterus, using a blunt needle. Each female was given 250 IU hCG intravenously at the time of surgery to induce ovulation. One of the 2 ferrets became pregnant and delivered at least 2 offspring, but cannibalized the litter soon after birth.

References

Blackshaw, A., and G. Salisbury. 1957. Factors influencing metabolic activity of bull spermatozoa. II. Cold-shock and its prevention. *J. Diary Sci.* 40:1099–106.

Chang, M. C., and R. Yanagimachi. 1963. Fertilization of ferret ova by depositing of epididymal sperm into the ovarian capsule with special reference to the fertilizable life of the ova and the capacitation of sperm. *J. Exp. Zool.* 154:175–87.

Curry, P. T., et al. A comparison of sperm morphology and silver nitrate staining characteristics in the domestic ferret and black-footed ferret. *Gamete Research.* 22:27–36.

Douzou, P. 1977. *Cryobiochemistry: An introduction.* New York: Academic Press.

Dresser, B. L., et al. 1987. Artificial insemination and embryo transfer in the felidae. In *Tigers of the World,* ed., R. L. Tilson and U. S. Seal. Park Ridge: Noyes.

Drevius, L. O. 1972. The permeability of bull spermatozoa to water polyhydric alcohols and univalent anions and the effects of the anions upon the kinetic activity of spermatozoa and sperm models. *J. Reprod. Fert.* 28:41–54.

Fougner, J. A., et al. 1973. Intrauterine insemination with frozen semen in the Blue Fox. *Nord. Vet.-med.* 25:144–49.

Kitchin, R. M., et al. 1988. Comparison of semen sperm content and sperm motility of European, Siberian and black-footed ferrets. *J. Androl.* 9:40.

Lindemann, C. B., et al. 1981. A comparative study of the effects of freezing and frozen storage on intact and demembranated bull spermatozoa. *Cryobiol.* 19:20–28.

Makler, A. 1978. A new chamber for rapid sperm count and motility estimation. *Fert. Steril.* 30:313–18.

Moreland, A. F., and C. Glaser. 1985. Evaluation of ketamine, ketamine-xylazine and ketamine-diazepam anesthesia in the ferret. *Lab. Animal Sci.* 35:287–90.

Neal, J., et al. 1977. Reproduction in the male ferret: Gonadal activity during the annual cycle; recrudescence and maturation. *Biol. Reprod.* 17:380–85.

Pilikian, S., et al. 1982. Effects of various concentrations of glycerol on post-thaw motility and velocity of human spermatozoa. *Cryobiol.* 19:147–53.

Seager, S. W. J. 1983. The breeding of captive wild species by artificial methods. *Zoo Biol.* 2:235–39.

Seal, U. S., and T. Foose. 1983. Development of a masterplan for captive propagation of Siberian Tigers in North American zoos. *Zoo Biol.* 2:241–44.

Shump, A. U., et al. 1976a. Semen volume and sperm concentration of the ferret (*Mustela putorius*). *Lab. Animal Sci.* 26:913–16.

Shump, A. U., et al. 1976b. Effect of electro-ejaculation on semen characteristics in the ferret (*Mustela putorius*). *Theriogen.* 7:83–87.

Watson, P. F. 1981. The effects of cold shock on sperm cell membranes. In *Effects of low temperature on biological membranes,* ed., G. J. Morris and A. Clarke. New York: Academic Press.

13

ROBERT W. ATHERTON, MONTE STRALEY, PATRICK CURRY, RICHARD
SLAUGHTER, WARREN BURGESS, AND ROBERT M. KITCHIN

A Review of Domestic Ferret Reproduction

Domestic ferrets (*Mustela putorius furo*) born in spring or summer are reproductively mature by the following spring. Testes of the mature ferret display full spermatogenic activity from March to July with regression occurring in August and September (Allanson 1931; Ishida 1968). After a period of quiescence, preparation for the next breeding season begins in December. There is a high correlation between testis size and plasma testosterone levels, the values of which gradually increase in December and January with a peak between April and June followed by a decline in July and August. Histological data also correlate with the seasonal variation in testosterone. In November, seminiferous tubules contain spermatogonia, Sertoli cells, and primary spermatocytes arrested in meiosis I. In December and January, new spermatogonia appear, and primary spermatocytes resume division. There are, however, no spermatids. By the end of February or March, all stages of spermatogenesis are present, and the epididymis contains sperm (Neal et al. 1977).

Luteinizing hormone-releasing hormone (LH-RH) causes an increase in luteinizing hormone (LH) and follicle stimulating hormone (FSH) secretion in males. Both FSH and LH appear to be secreted in an episodic manner with plasma concentrations increasing after gonadectomy. In males, an increase in plasma FSH is associated with an increase in plasma testosterone and an increase in testis size. Diurnal rhythms of LH appear to synchronize the episodic secretion of testosterone. Plasma FSH is low in mature aspermatogenic males, but in the regressed state the male retains the capability to respond to exogenous LH (Donovan and ter Haar 1977; Neal et al. 1977). Immature testes show little response to LH or FSH, indicating that some change must occur in the testis before maturation can proceed (Neal and Murphy 1977).

In males, seasonal variation of circulating testosterone and resulting variability in reproductive status are dependent on the hypothalamic-pituitary axis for control. There may be several factors that influence this control mechanism, as in other mammals; but probably the most important factor is photoperiod (Reiger and Murphy 1977). In male ferrets, short days stimulate reproductive activity while long days inhibit reproduction (Rust and Shackelford 1969).

Ovulation

In the ferret, ovulation occurs approximately 30 hours after coitus (Robinson 1918; Hammond and Walton 1934) and can sometimes occur as soon as 16 minutes to 3 hours after coitus. After injecting immature ferrets subcutaneously 4 to 5 times during a 2- to 4-day interval with 40 to 50 IU of pregnant mare serum (PMS), oocytes are found in the ovarian capsule 32 hours after a final 200 IU injection (Chang 1950). In addition, a single intraperitoneal administration of 90 IU of human chorionic gonadotropin (hCG) induced ovulation in 24 to 36 hours (Chang 1969).

In general, the ovum cytoplasm is evenly stained. Cellular extensions of corona radiata cells penetrate the zona pellucida. Relative age of an ovum can be determined by its morphology. The youngest ova have closely attached corona radiata cells, while older oocytes have a less intimate association with the corona radiata (Chang 1950). At 25 to 34 hours after ovulation, the second maturation spindle of the unfertilized ova is recognizable. At 41 hours after ovulation, the maturation spindle is irregular in shape. By 57 to 78 hours the first polar body divides, followed by degeneration of the unfertilized ovum (Chang and Yanagimachi 1963). Parthenogenetic cleavage occurs at a frequency of 43% in immature ferrets (Chang 1950) and 60% in mature ferrets (Chang 1957) but implantation does not occur.

Preliminary mating of estrous females to vasectomized males (to induce ovulation) followed by fertile matings at specific intervals seldom results in the fertilization of ferret ova older than 30 hours (Hammond and Walton 1934; Chang and Yanagimachi 1963). In hormonally ovulated ferrets, the number of ova fertilized by epididymal sperm injected into the ovarian capsule, which is continuous with the oviduct (Chang 1957), did not differ significantly within the first 24 hours following ovulation. After this 24-hour period, however, the percentage of fertilized ova decreased from 60 to 30% (Chang and Yanagimachi 1963).

The number of sperm ejaculated by ferrets is relatively low; only 10 to 20 million enter the female reproductive tract at ejaculation (Chang

1965). In contrast, 120 million sperm are ejaculated by rabbits (Marcirone and Walton 1938); and 4,250 million sperm are ejaculated by the ram (Chang 1946).

Three hours after coitus, motile sperm are found in the upper uterine horn, but not yet in the ovarian capsule—the site of fertilization in the ferret (Hammond and Walton 1934). After 6 hours, motile sperm are found in the ovarian capsule; but after 36 hours, they are absent from the capsule, although motile sperm may still be present in the uterus (Chang 1965).

Sperm apparently are capable of fertilization up to 126 hours post-intromission (Chang 1965); however, the rate of fertilization is time dependent. It can be demonstrated that 73% of eggs are fertilized if cauda epididymal sperm are injected into the uterine lumen 30 hours prior to ovulation. If ovulation occurs 42 to 60 hours after sperm deposition, however, the fertilization rate drops to 40%, while sperm in the tract for 78 to 126 hours are able to fertilize only 25% of the ova. Recent studies demonstrate that interrupted copulation reduces the percentage of successful fertilization in domestic ferrets (Miller and Anderson 1989).

Ferret sperm, like that of most mammals, prior to fertilization must first undergo an ill-defined process termed capacitation (Chang 1984), which occurs between 3.5 and 11.5 hours after sperm enter the female reproductive tract (Chang and Yanagimachi 1963).

Advances in Reproductive Technology

Attempts to artificially inseminate ferrets by mincing the epididymis in Tyrode's solution and vaginally depositing the solution prior to mating with sterile males has been reported (Hammond and Walton 1934). No pregnancies were obtained with this procedure, although sperm were introduced in a variety of concentrations. The protracted coitus in ferrets may function as a vaginal plug (Chang 1965), that is, the penis prevents sperm from escaping from the vagina prior to passage through the cervix.

Fertilized ova (as determined by penetration of the egg by sperm) can be obtained by surgical artificial insemination. The procedure consists of injecting cauda epididymal sperm into the ovarian capsule at various time intervals (Chang and Yanagimachi 1963). If deposition of sperm occurs between 6 hours prior to ovulation and 12 hours after ovulation, 54 to 66% of the ova are fertilized. Only 23% of the ova are fertilized if sperm deposition occurs 24 hours postovulation, whereas the proportion of fertilized ova drops to 14% at 36 hours postovulation. No successful vaginal artificial insemination attempts have been reported with electroejaculated ferret sperm (Atherton and Wildt, unpubl. data).

Ova fertilized by artificial insemination have been successfully transplanted from one ferret to another. This procedure required that pseudopregnant recipients and donor embryos be age-synchronized. Experiments were performed up to 8 days postfertilization. Although the most successful transfers occurred between eggs and uteri of the same age, there was an exception. Eight-day-old fertilized eggs were rescued at a higher rate when transferred to a 6-day-old uterus than to an 8-day-old uterus. In this case, however, only 29% of the transferred blastocysts developed into live fetuses. The best rescue rate was in the transfer of 4-day-old fertilized eggs to a 4-day-old pseudopregnant uterus. Over 40% of the eggs developed into live fetuses. Although live fetuses were produced, no pregnancy was allowed to go to term (Chang 1969).

Cryopreservation of sperm depends on membrane protection. Freezing and thawing cause changes in osmotic potential of the solution surrounding cells that can potentially denature membrane-associated proteins or rupture membranes (Meryman 1968). In addition, the growth of intracellular ice crystals is likely to cause disruption of the structural elements within cells (Mazur 1966). Changes in the structure of proteins and lipids due to cell shrinkage prior to freezing also cause cell damage that can lead to cell death (Litvan 1972). There is also evidence that during freezing and thawing of cells, some constituents leak through the membrane to the cell exterior (Graham and Pace 1967). Morphologic studies show that the acrosome in particular is affected by cold shock (Watson 1981). Cryoprotective agents (CPA) at high concentrations may destabilize membrane bilayers by dissolving phospholipids. At moderate concentrations, however, they prevent the damaging effects of phospholipid dehydration upon freezing. Proteins appear to stabilize bilayers by providing increased hydration at the membrane surface and by causing additional hydrophobic binding in the membrane interior (Douzou 1977; Graham and Pace 1967).

Chilling injury is damage caused by reduction in temperature, with no water-ice phase change in the system. This is also referred to as cold shock or thermal injury. The rate of cooling is an important variable in determining the amount of cold-shock injury. Freezing injury is damage incurred during a temperature reduction severe enough to cause a water-ice phase change in the system (Watson 1981). During freezing, ice formation increases the solute concentration in the unfrozen portion of an aqueous solution. Ice formation, which is important intracellularly, depends on the water conductivity of the cell membrane and rate of cooling (Graham et al. 1972) and may lead to a destructive rise in salt concentration and denaturation of lipid-protein complexes (Jeyendran et al. 1979). A CPA can prevent destabilization by reducing loss of water and possibly by lowering the liquid crystal-gel transition temperature within the cell (Jeyendran and Graham 1981; Jeyendran et al. 1985).

Materials that fail to crystallize form glasses at low temperatures. Successful cryopreservation appears to force extracellular solutions into a glassy state rather than allowing formation of ice during cooling. One problem of vitrification is that the solutions invariably devitrify, that is, form ice crystals during subsequent rewarming to room temperature, even though crystallization was avoidable during cooling. The devitrification process becomes less extensive at higher heating rates (Mac-Farlane 1986; Boutron 1986; Brower et al. 1981).

Egg yolk has long been known to provide protection to ram and bull sperm undergoing cold shock and exposure to low temperature. Several other substances, such as milk and milk proteins, bovine and egg albumin (Choong and Wales 1962), also provide a measure of protection against cold shock. The active fraction of egg yolk is the low-density lipoprotein (Pace and Graham 1974). The protective action may be due to its solubility in or binding to the sperm surface (Watson 1981).

Glycerol is a common cryopreservative agent, presumably because it is a normal intermediary product of lipid metabolism. The protective effect appears to be due to the stabilization of membrane surfaces (Douzou 1977). Egg yolk and low-molecular-weight compounds, like glycerol, probably interact synergistically with each other or with the sperm membrane during freezing. For example, bull semen is reported to have a macromolecular interaction between egg yolk and seminal plasma proteins that may be responsible for coating the sperm surface and thus aiding in cryoprotection (Jeyendran et al. 1979). A low-molecular-weight compound such as glycerol may promote formation of the above interaction (Jeyendran and Graham 1980).

It appears that damage to the plasma membrane of sperm is the major cause of motility loss when bull sperm are thawed. The loss of motility resulting when intact sperm are frozen is apparently due to a damaged plasma membrane. However, demembraned bull sperm may be kept motile with 2 mM of ATP; thus, the motile apparatus of demembraned bull sperm remains functional upon thawing. When observed in electron micrographs, intact sperm cooled at 180°C per hour showed numerous breaks in the plasma membrane as well as a disorganization of the mitochondrial interior structures. Such membrane damage appears to stop motility, primarily because endogenous ATP is permitted to leak out of the cells. When exogenous Mg^{+2} and ATP are supplied, sperm motility resumes (Lindemann et al. 1981).

Cold-shock variability between species and individuals within a species is well documented (Pursel et al. 1973) and for bull spermatozoa at least, variability also exists between spermatozoa within an ejaculate (Choong and Wales 1962). Approximately 25% of bull sperm were found to be resistant to repeated cold shock. Thus, an individual cell may have an all-or-none response, but variations in susceptibility exist between cells in a population (Watson 1981).

Cooling rates from body temperature to 4°C are also important for sperm viability. Cooling rates of 20°C per hour may be used above 15 to 20°C but should be reduced to approximately 10°C per hour below 15 to 20°C to maintain viability of ram and bull sperm. The necessity of a slow cooling rate to maintain viability as the temperature is decreased suggests that lipid phase transitions are involved, especially at lower temperatures (Watson 1981). Spermatozoa that have been cooled slowly, thus avoiding cold shock, are affected by continued exposure to low temperature. Such exposure produces changes in the plasma membrane, acrosome, and mitochondria, and if a CPA is not included, loss of motility and fertility result (Franks 1985).

Examples of sperm cryopreservation of zoological interest are found in cats and foxes. In domestic cats, 10.7% of artificial inseminations were successful with the following protocol and extender: semen was mixed with 200 μl of sterile saline (0.9%) and 200 μl of diluent at room temperature (22 to 23°C) at approximately 1 : 1 with an extender of deionized water with 20% EY, 11% (w/v) lactose, and 4% (v/v) glycerol with 1,000 μg of streptomycin and 1,000 IU per ml of Penicillin G. Samples were pelleted on dry ice and placed in liquid nitrogen. Thawing took place in 0.154 M NaCl at 37°C. Some differences were exhibited in sperm viability depending on use of an artificial vagina or electroejaculation. The artificial vagina produced 83% fresh motility and 71% post-thaw, while electroejaculation produced 70 and 54% respectively (Platz et al. 1978).

Fox semen was collected by digital manipulation and diluted 1 : 3 to 1 : 5 in a tris-fructose-citric acid-extender containing 8% glycerol and 20% egg yolk, giving final sperm concentrations of 150 to 200 million per ml. A 3-hour cooling period was used to reach 4 to 5°C and to equilibrate. The samples were then placed in straws and frozen in liquid nitrogen vapors. Thawing the samples at 75°C for 9.5 seconds gave 55 to 65% motile sperm. Nine of 11 vixens artificially inseminated became pregnant, approximately the same success as obtained in natural breeding (Fougner et al. 1973; Aamdal et al. 1978).

Although the domestic ferret is well studied from an anatomical and endocrine viewpoint, both a photoperiod regime that maximizes sperm production and hormonal manipulation of ovulation need to be established for efficient studies with Siberian and black-footed ferrets. In addition, a nonsurgical procedure to artificially inseminate ferrets needs to be developed.

Cryopreservation of sperm, while technically rather simple, necessitates an understanding of complex cryobiology issues to effectively design a new protocol. In ferrets, this is especially difficult because the small ejaculate volume hampers experimental design. Furthermore, specific photoperiod requirements for active spermatogenesis temporarily limit studies. Nevertheless, cryopreservation of domestic ferret sperm is possible. It must be determined if these protocols are valid for Siberian ferrets and black-footed ferrets and if pregnancy can be achieved by artificial insemination with both freshly collected and post-thaw ejaculates.

References

Aamdal, J., et al. 1978. Artificial insemination in foxes. *Symp. Zool. Soc. Lond.* 43:241–48.

Allanson, M. 1931. The reproductive processes of certain mammals. III. The reproductive cycle of the male ferret. *Proc. Roy. Soc.*, series B, 110:296–312.

Brower, W. E., et al. 1981. An hypothesis for survival of spermatozoa via encapsulation during plane front freezing. *Cryobiol.* 18:277–91.

Boutron, P. 1986. Comparison with the theory of the kinetics and extent of ice crystallization and of the glass forming tendency in aqueous cryoprotective solutions. *Cryobiol.* 23:88–102.

Chang, M. C. 1946. The sperm production of adult rams in relation to frequency of semen collection. *J. Agri. Sci.* 35:243–46.

———. 1950. Cleavage of unfertilized ova in immature ferrets. *Anat. Rec.* 108:31–43.

———. 1957. Natural occurrence and artificial induction of parthenogenetic cleavage of ferret ova. *Anat. Rec.* 128:187–99.

———. 1965. Fertilizing life of ferret sperm in the female tract. *J. Exp. Zool.* 158:87–99.

———. 1969. Development of transferred ferret eggs in relation to the age of the corpora lutea. *J. Exp. Zool.* 171:459–64.

———. 1984. The meaning of sperm capacitation: A historical perspective. *J. Androl.* 5:45–50.

Chang, M. C., and R. Yanagimachi. 1963. Fertilization of ferret ova by deposition of epididymal sperm into the ovarian capsule with special reference to the fertilizable life of ova and capacitation of sperm. *J. Exp. Zool.* 154:175–87.

Choong, C. H., and R. G. Wales. 1962. The effect of cold shock on spermatozoa. *Aust. J. Biol. Sci.* 15:543–51.

Donovan, B. T., and M. B. ter Harr. 1977. Stimulation of the hypothalamus by FSH and LH secretion in the ferret. *Neuroendocrin.* 23:268–78.

Douzou, P. 1977. *Cryobiochemistry: An introduction.* New York: Academic Press.

Fougner, J., et al. 1973. Intrauterine insemination with frozen semen in the blue fox. *Nord. Vet.-med.* 25:144–49.

Franks, F. 1985. *Biophysics and biochemistry at low temperatures.* New York: Cambridge Univ. Press.

Graham, E. F., and M. M. Pace. 1967. Some biochemical changes in spermatozoa due to freezing. *Cryobiol.* 4:75–84.

Graham, E. F., et al. 1972. Effects of some zweitterion buffers on the freezing and storage of spermatozoa. *J. Diary Sci. Bull.* 55:372.

Hammond, J., and A. Walton. 1934. Notes on ovulation and fertilization in the ferret. *J. Exp. Biol.* 11:307–19.

Ishida, K. 1968. Age and seasonal changes in the testis of the ferret. *Arch. Histol. Jap.* 29:193–205.

Jeyendran, R. S., and E. F. Graham. 1980. An evaluation of cryoprotective compounds on bovine spermatozoa. *Cryobiol.* 17:458–64.

———. 1981. Effects of cooling and freezing on pH of semen extender. *Cryobiol.* 19:16–19.

Jeyendran, R. S., et al. 1979. Interaction of seminal plasma proteins with egg yolk during cryopreservation of bovine semen. *Proc. 71st Ann. Am. Soc. Anim. Sci.* Abst. 405:305.

Jeyendran, R. S., et al. 1985. Nonbeneficial effects of glycerol on the oocyte penetrating capacity of cryopreserved and incubated human spermatozoa. *Cryobiol.* 22:434–37.

Lindemann, C. B., et al. 1981. A comparative study of the effects of freezing and frozen storage on intact and demembraned bull spermatozoa. *Cryobiol.* 19:20–28.

Litvan, G. G. 1972. Mechanisms of cryoinjury in biological systems. *Cryobiol.* 9:182–91.

MacFarlane, D. R. 1986. Devitrification in glass forming aqueous solutions. *Cryobiol.* 23:230–44.

Marcirone, C., and A. Walton. 1938. Fecundity as determined by "Dummy Matings." *J. Agri. Sci.* 122.

Mazur, P. 1966. Physical and chemical basis of injury in single celled microorganisms subjected to freezing and thawing. *Cryobiology*, ed. H. T. Meryman. New York: Academic Press.

Meryman, H. T. 1968. Modified model for the mechanism of freezing in erythrocytes. *Nature* 218:333–36.

Miller, B. J., and S. H. Anderson. 1989. Precopulatory behavior of induced oestrous and natural estrous in domestic ferrets. *Animal Behavior.* Submitted.

Neal, J., and B. D. Murphy. 1977. Response of immature, mature non-breeding and mature breeding ferret testis to exogenous LH stimulation. *Biol. Reprod.* 16:244–48.

Neal, J., et al. 1977. Reproduction in the male ferret: Gonadal activity during the annual cycle; recrudescence and maturation. *Biol. Reprod.* 17:380–85.

Pace, M. M., and E. F. Graham. 1974. Components in egg yolk which protect bovine spermatozoa during freezing. *J. Anim. Sci.* 39:1144–49.

Platz, C. C., et al. 1978. Pregnancy in the domestic cat after artificial insemination with previously frozen spermatozoa. *J. Reprod. Fert.* 52:279–82.

Pursel, V. G., et al. 1973. Effect of dilution, seminal plasma and incubation period on cold shock susceptibility of boar spermatozoa. *J. Anim. Sci.* 37:528–31.

Reiger, D., and B. D. Murphy. 1977. Episodic fluctuation in plasma testosterone in male ferret during the breeding season. *J. Reprod. Fert.* 51:511–14.

Robinson, A. 1918. The formation, rupture, and closure of ovarian follicles in ferrets and ferret-polecat hybrids and some associated phenomena. *Trans. Roy. Soc. Edinburgh* 52:303–63.

Rust, C., and R. M. Shackelford. 1969. Effect of blinding on reproductive and pelage cycles in the ferret. *J. Exp. Zool.* 171:443–50.

Watson, P. F. 1981. The effects of cold shock on sperm cell membranes. In *Effects of low temperatures on biological membranes*, ed. G. J. Morris and A. Clark, 189–218. New York: Academic Press.

PART FOUR

Management and Conservation

14

BRIAN P. COLE

Recovery Planning for Endangered and Threatened Species

The Endangered Species Act (ESA) as amended in 1982 (Pub. L. 97–304) directs the secretaries of interior and commerce to develop a list of species that are in danger of extinction and to carry out programs for the conservation of listed species. The secretary of commerce has delegated this authority to the National Marine Fisheries Service (NMFS), which is responsible for most marine species. The secretary of the interior has delegated to the Fish and Wildlife Service (FWS) the authority for certain marine species (when on shore) and for all other listed species.

Under the ESA, species may be listed as threatened or endangered. Endangered species are those in danger of extinction throughout all or a significant portion of their range. Threatened species are those likely to become endangered within the foreseeable future.

In 1978, Section 4(f) of the ESA was amended to require the development of recovery plans and a priority system for recovery of listed species:

> Recovery Plans—The Secretary shall develop and implement plans (hereinafter in this subsection referred to as 'recovery plans') for the conservation and survival of endangered species and threatened species listed pursuant to this section unless he finds that such a plan will not promote the conservation of the species. The Secretary, in developing and implementing recovery plans (1) shall, to the maximum extent practicable, give priority to those endangered species or threatened species most likely to benefit from such plans, particularly those species that are, or may be, in conflict with construction or other developmental projects or other forms of economic activity, and (2) may procure the services of appropriate public and private agencies and institutions, and other qualified persons. Recovery teams appointed pursuant to this subsection shall not be subject to the Federal Advisory Commission Act.

Table 14.1.   U.S. Listed Species with
Approved Recovery Plans by Taxonomic
Group (as of August 1988)

Taxon	No. of Species with Approved Plans
Mammals	25
Birds	60
Reptiles	22
Amphibians	5
Fish	45
Snails	7
Clams	22
Crustaceans	2
Insects	12
Plants	71
Total	271

Note: Listed species worldwide = 1001; listed
species for U.S. = 504; approved recovery
plans for U.S. = 230.

The act further requires that other federal agencies assist the FWS
and NMFS in efforts to conserve listed species. Section 7(a)(1) states,
"All other Federal agencies shall, in consultation with and with the
assistance of the Secretary, utilize their authorities in the furtherance of
the purposes of this Act by carrying out programs for the conservation
of endangered species and threatened species listed pursuant to Section
4 of this Act."

Some statistics will help to illustrate the magnitude of the task. Currently, over 1,000 species worldwide are classified on the federal list as
endangered or threatened. Of these, over 500 inhabit the United States
or its territories. Available data indicate that about 1,000 additional
species warrant designation as threatened or endangered; another
2,000 candidate species may warrant listing, but sufficient data are not
yet available. About 50 such candidates are added to the list each year.
The FWS focuses recovery plan development and implementation on
listed species in the United States and its territories. To date, approximately 230 recovery plans have been approved, covering about 270
U.S. listed species (Table 14.1).

Priorities

In 1983, the FWS published in the *Federal Register* (Vol. 48, No. 184,
4309–43105) a priority system for listing and recovering endangered

and threatened species, which ensures that limited resources are focused on the highest priority tasks. Priorities are based on the degree of threat facing a species, taxonomic uniqueness, recovery potential, and whether it is likely to encounter conflicts with development (Table 14.2). The taxonomic criterion is intended to give higher priority to species representing distinctive gene pools in that loss of the most genetically distinct taxa is of greater significance than loss of those less so. A separate task priority system, used in conjunction with the species priorities, assigns a priority of from 1 to 3 to recovery tasks as follows:

Priority 1. An action that must be taken to prevent extinction or to prevent the species from declining irreversibly in the foreseeable future;

Priority 2. An action that must be taken to prevent a significant decline in species population/habitat quality, or some other significant negative impact short of extinction;

Priority 3. All other actions necessary to provide for full recovery of the species.

The species and task priority systems provide a two-tiered approach to ensuring that limited resources are effectively and efficiently allo-

Table 14.2. Species Recovery Priorities of the Fish and Wildlife Service

Degree of Threat	Recovery Potential	Taxonomy	Priority	Conflict
High	High	Monotypic genus	1	1/1C[a]
	High	Species	2	2/2C
	High	Subspecies	3	3/3C
	Low	Monotypic genus	4	4/4C
	Low	Species	5	5/5C
	Low	Subspecies	6	6/6C
Moderate	High	Monotypic genus	7	7/7C
	High	Species	8	8/8C
	High	Subspecies	9	9/9C
	Low	Monotypic genus	10	10/10C
	Low	Species	11	11/11C
	Low	Subspecies	12	12/12C
Low	High	Monotypic genus	13	13/13C
	High	Species	14	14/14C
	High	Subspecies	15	15/15C
	Low	Monotypic genus	16	16/16C
	Low	Species	17	17/17C
	Low	Subspecies	18	18/18C

[a]C = conflict with development.

cated. Development and implementation of recovery plans, budget development, and evaluation of funding proposals are all guided by this
priority system.

Recovery Planning

The FWS recovery plans are intended to guide the conservation programs of federal, state, and local agencies and other organizations. The
purpose of a recovery plan is to identify the tasks necessary to make a
species a viable, self-sustaining component of its ecosystem and no longer in need of the protection of the ESA. This would normally lead to
"delisting." Due to the critical status of many species, however, recovery
plans often focus on the more immediate goal of preventing extinction
(the California condor, black-footed ferret, and Puerto Rican parrot).
An intermediate objective is frequently provided which would result in
"downlisting" from endangered to threatened status.

Typically, each listed species has its own recovery plan, but in some
instances one plan will cover a group of species in similar ecosystems and
with common threats. Conversely, some species are so wide-ranging (the
bald eagle and peregrine falcon) that more than one plan is required to
accommodate divergent habitat requirements. The length and complexity of a plan varies with the geographical distribution of the species,
the complexity of problems it faces, and the number of organizations or
agencies involved.

The FWS developed *Recovery Planning Guidelines* (FWS 1985) to ensure consistency and high quality in Service-approved recovery plans.
The guidelines specify the format and general content of these plans,
each of which is divided into 3 major sections:

> Part I—Introduction. Contains background information of habitat require-
> ments, past and current distribution, status, taxonomy, limiting factors, and
> conservation efforts;
>
> Part II—Recovery. States a quantitative recovery objective (where possible)
> that would lead to reclassification or de-listing and outlines all tasks neces-
> sary to achieve the objective;
>
> Part III—Implementation Schedule. Assigns task priorities, identifies re-
> sponsible agencies, and estimates the cost and timing of tasks.

The FWS regional directors are responsible for preparing recovery
plans for species within their region; for multiregional species, a lead
region is designated. Generally, plans are prepared first for species with

the highest recovery priorities (Table 14.2). The regional director can have a plan prepared by a volunteer recovery team, a state or federal agency, conservation organization, knowledgeable individuals (by contract or voluntarily), or FWS personnel.

The method used to prepare a plan depends on the range of the species, the complexity of the recovery effort, the number of responsible agencies, and the expertise of available personnel. When the FWS first began this work in 1975, development of most plans was assigned to recovery teams organized by the FWS and consisting of experts on the species from the public and private sectors. As more species have been listed and the FWS has gained experience, more plans have been developed by contract or by FWS personnel. When adequate expertise is available, contracts generally result in faster plan development.

Recovery plan development takes about 1 year. Prior to approval, recovery plans may go through three separate reviews involving FWS personnel in the regions and the Washington office, federal and state agencies, conservation groups, and species experts. Originally, the director of the FWS approved all plans, but now most are approved by the regional directors unless multiregional or nationally significant species are involved (the grizzly bear, bald eagle, and whooping crane). Each recovery plan is reviewed annually and is revised as warranted by new information or changes in the status of a species.

Once a recovery plan is approved, a limited number are printed for distribution to FWS offices and other agencies, organizations, and individuals involved in implementation. Photocopies and microfiche copies of approved plans are available through the Fish and Wildlife Reference Service (5430 Grosvenor Lane, Suite 110, Bethesda, MD 20814; [301] 492-6403/[800] 582-3421).

It should be emphasized that recovery plans are intended to assist the FWS and other organizations involved in the conservation of listed species. No agency or organization is obligated to carry out tasks listed in a recovery plan. The FWS is well aware of the legal, policy, fiscal, and political constraints that determine an agency's ability to implement programs. By using recovery plans as guides, however, agencies can ensure that their conservation efforts are focused on the highest priority actions.

FWS Ferret Recovery Efforts

The FWS first became involved in efforts to conserve black-footed ferrets in 1965, when the assistant secretary of the interior for fish, wildlife, and parks recognized this species as endangered. He also acknowledged the dual role of the department in conserving ferrets and controlling prairie dogs. In 1965, the service began systematic surveys for ferrets on the Pine Ridge Indian Reservation in South Dakota (Schroeder 1985).

In 1966, Congress passed the Endangered Species Preservation Act (Pub. L. 89–699) which called for developing a list of native fish and wildlife threatened with extinction. Although this act recognized the necessity of preventing the extinction of native fauna and directed the secretary of the interior to carry out programs for their protection and restoration, the only significant authority provided was for land acquisition for endangered species. No federal prohibitions on taking, interstate commerce, or habitat destruction were included (Bean 1983). In 1967, the black-footed ferret was listed under this act.

Also in 1966, the Patuxent Wildlife Research Center (PWRC) in Laurel, Maryland, recruited a biologist to study the black-footed ferret at a research facility in Rapid City, South Dakota (Schroeder, telephone conversation, 1986). (The Rapid City Field Station focused on the study of black-footed ferrets in that state until it was closed in 1980.) In 1968, PWRC began surrogate studies with the European ferret (*Mustela putorius*), which were expanded with the acquisition of 46 Siberian ferrets (*Mustela eversmanni*) from the Soviet Union in 1975 (Carpenter 1985).

Congress recognized the deficiencies in the 1966 Endangered Species Preservation Act and attempted to remedy them in 1969 by passing the Endangered Species Conservation Act (Pub. L. 91–135). That act, among other things, called for developing a list of species threatened with worldwide extinction, expanded the land acquisition authority of the 1966 act, and prohibited import of listed species into the United States (Bean 1983). The black-footed ferret was placed on the list promulgated under this act in 1970.

In 1971, 6 of the wild South Dakota ferrets were captured (four females and two males) and transported to PWRC in Laurel, Maryland, to begin a captive-propagation program, the purpose of which was to study the ferret and to produce stock for restoring or bolstering wild populations. The 4 females died shortly after capture from vaccine-induced canine distemper. Additional South Dakota ferrets were captured in 1972 (1 female) and in 1973 (1 male and 1 female); all were transported to PWRC. Although the surrogate European and Siberian ferrets successfully produced a large number of kits, only 1 male and 1 female among the black-footed ferrets mated successfully (once in 1976 and once in 1977). The female produced 5 young each year: 4 in each litter were stillborn, and the 5th died within 1 to 2 days. The last captive black-footed ferret at PWRC died in 1979 (Carpenter 1985).

In 1973, Congress passed the Endangered Species Act (Pub. L. 93–205)—a much stronger and more comprehensive statute than either the 1966 or 1969 acts—which developed a worldwide list of threatened or endangered animals and plants. The black-footed ferret was included on this list.

In 1974, the FWS established a black-footed ferret recovery team to develop a recovery plan. The team included representatives from the FWS; National Audubon Society; South Dakota Department of Wildlife, Parks, and Forests; South Dakota Cooperative Wildlife Research Unit; National Park Service; Bureau of Land Management; and Forest Service. On June 14, 1978, the FWS approved the *Black-footed Ferret Recovery Plan,* which had as its primary objective to "maintain at least one wild self-sustaining population . . . in each State within the ferret's former range."

Because no ferrets were known to exist in the wild at the time the recovery plan was approved, FWS recovery efforts from 1978 to 1981 were limited to surveys for ferrets by the Denver Wildlife Research Center (DWRC). Most of these surveys were funded by the private sector or by federal land management agencies that had proposed projects (coal mining, pipelines, and so on) affecting potential ferret habitats. Since Section 7 of the ESA requires federal agencies to insure that actions they authorize, fund, or carry out are not likely to jeopardize the continued existence of listed species, ferret surveys are required prior to federal approval of many projects in potential ferret habitat. The FWS has developed ferret survey guidelines (FWS 1986) to assist other federal agencies and developers in complying with Section 7 of the ESA.

On September 26, 1981, a ranch dog killed a black-footed ferret near Meeteetse, Wyoming. Subsequent searches of prairie dog towns in the area led to the discovery of an apparently healthy population of black-footed ferrets. Since that discovery, the FWS, the state of Wyoming, and private organizations have cooperated in efforts to study the Meeteetse population.

The Meeteetse studies have focused on the ecological requirements, population dynamics, and movements of the ferrets. In 1981, DWRC personnel captured 1 ferret, equipped it with a radio transmitter, and monitored its movements for 2 weeks. In 1982, radio-tracking studies were continued on 6 young ferrets. In 1983, various marking techniques were tested and, in conjunction with continued radio tracking and snow tracking, yielded information on movement, home range, activity patterns, and mortality (Fagerstone and Biggins 1986).

In 1984, new radio-collars for ferrets were tested, mark-recapture studies were conducted, studies of the prairie dog prey base were begun, and various predators were radio-collared. In the summer of 1984, mark-recapture studies gave a population estimate of 129 ferrets, and spotlighting revealed 126 ferrets.

During 1985, prairie dog studies were continued, and small mammal trapping was begun. Monitoring for sylvatic plague was initiated, and fleas from 4 of 6 prairie dog colonies tested positive. To reduce the spread of plague, prairie dog burrows were dusted with carbaryl to kill the fleas, but surveys showed a 22% decline in acreage occupied by prairie dogs. Spotlighting turned up 58 ferrets in July and August of 1985 (Fagerstone and Biggins 1986).

In September and October 1985, 6 ferrets were brought into captivity. Soon afterward, 1 showed signs of canine distemper. All 6 captive animals eventually succumbed to the disease. Later in the fall of 1985, 6 more animals (2 immature males, 2 immature females, and 2 mature females) were captured, vaccinated against distemper, and held in isolation to prevent the possible spread of distemper. When the animals were determined to be free of distemper, they were moved to the state's temporary breeding facility at the Sybille Wildlife Center near Wheatland, Wyoming.

Spotlighting in the summer of 1986 revealed only 1 adult male, 2 adult females with litters, and 2 unclassified ferrets in the wild. The adult male and 3 juvenile females were captured for the captive-breeding program. In the fall of 1986, all of the remaining ferrets were also captured.

When the FWS approved the *Black-footed Ferret Recovery Plan* in 1978, no ferrets were known to exist in the wild. The discovery of the Meeteetse population and subsequent research findings necessitated a major revision of the recovery plan. A revised recovery plan was drafted in 1985, but the rapid decline of the ferret population and the initiation of the captive-breeding effort again made it obsolete. Another revision will be undertaken based on the results of the 1986 summer surveys and the captive-breeding program.

The FWS will continue to support the recovery of the black-footed ferret. Table 14.3 illustrates FWS funding of ferret recovery efforts since discovery of the Meeteetse population in 1981. It is apparent that the captive-breeding effort must proceed, along with development of reintroduction techniques. Surveys to locate additional populations and to investigate reported sightings of ferrets are necessary. Potential reintroduction sites must be identified. Additional research on ecology, husbandry, physiology, behavior, and reproduction are needed for both surrogate species and black-footed ferrets.

The combined efforts of federal and state agencies, conservation groups, private landowners, and the scientific community will be re-

Table 14.3.   Fish and Wildlife Service Funding History for the Black-Footed Ferret

Activity	Fiscal Year							Total
	1982	1983	1984	1985	1986	1987	1988	
Research	67.0[a]	67.0	104.5	103.5	116.0	216.0	210.0	884.0
Recovery planning	2.0	2.0	2.0	2.0	2.0	63.0	63.0	136.0
Off-service lands		12.0	25.5	25.4	218.0	24.0	64.0	368.9
Section 6		42.0	65.0	271.0	320.0	255.5	225.0	1178.5
Total	69.0	123.0	197.0	401.9	656.0	558.5	562.0	2567.4

[a]In thousands of dollars.

quired to turn the black-footed ferret back from the brink of extinction. Although success cannot be assured, the ESA provides us with the tools necessary to proceed and with the mandate that we try.

References

Bean, M. J. 1983. *The evolution of national wildlife law.* Environmental Defense Fund. New York: Praeger Publishers.

Carpenter, J. W. 1985. Captive breeding and management of black-footed ferrets. In *Black-footed ferret workshop proceedings,* 12.1–12.13. Cheyenne, Wyo.: Wyoming Game and Fish Department.

Fagerstone, K. A., and D. E. Biggins. 1986. Summary of black-footed ferret and related research conducted by the Denver Wildlife Research Center, 1981–1985. Appendix 1 to the draft revised *Black-footed ferret recovery plan.* Denver, Colo.: U.S. Fish and Wildlife Service.

Schroeder, M. H. 1985. U.S. Fish and Wildlife Service guidelines for black-footed ferret surveys. In *Black-footed ferret workshop proceedings,* 27.1–27.5. Cheyenne, Wyo.: Wyoming Game and Fish Department.

U.S. Fish and Wildlife Service. 1985. *Recovery planning guidelines.* Washington, D.C.: U.S. Fish and Wildlife Service.

———. 1986. *Black-footed ferret survey guidelines for compliance with the Endangered Species Act.* Denver, Colo., and Albuquerque, N. Mex.: U.S. Fish and Wildlife Service.

15

THOMAS J. FOOSE

Species Survival Plans: The Role of Captive Propagation in Conservation Strategies

Conservation strategies traditionally have placed primary attention on protection of natural habitats and, by extension, of their resident populations. Such actions are obviously necessary and desirable if the evolutionary and ecological integrity of a species is to be preserved. Such actions may not be sufficient, however, where degradation of habitat has greatly reduced and fragmented natural populations. Moreover, many wildlife species are decimated even before other elements of their natural habitat. In either case, the resultant small, wild populations become vulnerable to and possibly extinct by the kinds of genetic and demographic problems that have been elaborated in other chapters of this book.

As a consequence, species survival plans (SSPs) must be predicated on population biology and management. The objective of species survival plans is the preservation of wildlife both as species (gene pools) and as constituents of their ecosystems. Thus, species survival plans employ both *in situ* and *ex situ* methods, that is, both captive and wild populations. Finally, there are both biological and organizational components to species survival plans. This chapter will discuss these aspects, with special reference to species survival plans being developed with the involvement of the zoo community, and in relation to the black-footed ferret.

Captive and Wild Populations

Preservation of species and their gene pools need not and frequently can not depend solely on wild populations. Captive propagation can help. Indeed, for many species—the Arabian oryx (Stanley-Price 1986), the Przewalski horse (FAO 1986), the California condor (Crawford 1985a, 1985b), and perhaps the black-footed ferret—the natural habitat may become untenable for a time, and the wild populations will temporarily disappear. *Ex situ* captive populations must then sustain the species until it can be restored to a natural habitat. In general, both captive propagation and wild populations may be necessary for the survival of many species, including the black-footed ferret.

The primary purpose of captive propagation, however, is to reinforce, not replace, wild populations. This reinforcement can be both genetic and demographic. Genetically, captive-propagation programs can serve as a reservoir of genetic material that can be transfused periodically as new blood into "anemic" wild populations. Demographically, captive populations can also serve as a source of new stock to compensate for numerical deficiencies in the wild. In essence, captive propagation can significantly moderate stochastic perturbations that harass small wild populations. Small captive populations naturally will be susceptible to stochastic perturbations; but the level of genetic, demographic, veterinary, and nutritional management normally possible in captive situations will mollify these effects. The possibility of managing captive populations to maximize the genetically effective size (N_e) is of particular significance.

Wild populations, of course, are vital in assuring natural selection and thus in maintaining desirable wild traits, which eventually erode in captivity despite the best attempts to arrest long-term genetic change. Survival of the species as a component of natural ecosystems is the ultimate objective of all conservation efforts. But frequently, wild populations by themselves simply will not survive, no matter how heroic the conservation efforts.

Thus, optimal conservation strategies for endangered species like the black-footed ferret should incorporate both captive and wild populations that are interactively managed for mutual support as schematized in Figure 15.1 (Foose et al. 1985). Moreover, similar principles and methods of genetic and demographic management should be applied to small populations in the wild as well as in captivity. For example, the intensive management possible in captivity depends upon individual identification and data management. For species such as the black-footed ferret, whose wild populations are very small and whose individuals are markable and manipulable, a program of individual identification and management may be feasible and desirable. Indeed, such programs are already under way for several species—tigers (Smith et al. 1986) and black rhino (Western, pers. comm.). As Sullivan and Shaffer (1975), among others, have observed, wildlife management in the wild, as well as in captivity, is becoming somewhat of a "megazoo."

This vision of cooperative captive and wild programs is not without its

Captive Populations   Wild Populations

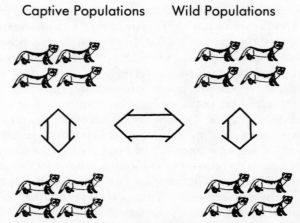

Figure 15.1. Interactively managed system of captive and wild populations in an overall conservation strategy for black-footed ferrets.

critics, many of whom frequently express skepticism about the feasibility of reintroducing captive stock into wild situations. Two major concerns are that animals produced in captivity will not be able to adapt to natural conditions and that reintroductions will be disruptive to remaining wild populations.

The first concern is especially significant for a predatory species like the black-footed ferret whose survival skills may be culturally transmitted. It will be impossible to determine the seriousness of this problem until reintroduction is attempted. (Apparently successful reintroductions of captive stock have occurred or are now occurring [Stanley-Price 1986; Kleiman and Beck 1986].) Even if readaptation is a difficult process, however, it can probably be accomplished as long as the captive population produces enough animals for reintroduction to sustain the losses as captive stock readapts.

The second problem of disruption to remnant wild populations does not currently apply to the black-footed ferret, but may become serious in the future if and when wild populations are reestablished. For other species that may experience such disruption, reproductive technology seems to provide the prospect of assistance. It appears entirely possible that genetic material can be transferred from captive to wild populations via artificially transplanted sperm or embryos. The technology for such operations is not yet available, but vigorous research is in progress (Dresser et al. 1986; Seal et al. 1986).

Reproductive technology may assist in yet another way. The "frozen zoo" (in the form of cryogenically preserved gametes or embryos) may provide the potential of expanding effective populations to thousands or millions of animals. This prospect does not propose that living animals be replaced, but once again, merely reinforced. Populations of living animals in both natural habitats and captive facilities must continue to be maintained, but also augmented by stored germinal tissue where technologically possible.

Zoos are organizing to perform this integral role in conservation strategies. In North America, the Species Survival Plan (SSP) has been established (AAZPA 1983; Foose 1983) and encompasses all zoos in the United States and Canada maintaining a particular species. Thirty-eight species have been designated for the SSP thus far (Table 15.1). It is expected that SSP programs will eventually exist for at least 1,000 taxa of mammals, birds, reptiles, and amphibians. An international system is evolving as zoos in other regions of the world develop similar programs. The Captive Breeding Specialist Group (CBSG) of IUCN is instrumental in this development.

For each species designated, an intensive program of demographic and genetic management is developed (Seal and Foose 1983a, 1983b).

Table 15.1. Taxa Designated for the AAZPA Species Survival Plan (SSP)

Amphibians	Mammals (*continued*)
Puerto Rican crested	Maned wolf
toad	Red wolf
Reptiles	Red panda
Chinese alligator	Siberian tiger
Radiated tortoise	Asian lion
Aruba Island rattlesnake	Snow leopard
Madagascar ground boa	Clouded leopard
Birds	Asian elephant
Bali mynah	Sumatran rhino
White-naped crane	Indian rhino
Andean condor	Black rhino
Humboldt's penguin	White rhino
Mammals	Grevy's zebra
Ruffed lemur	Asian wild horse
Black lemur	Chacoan peccary
Golden lion tamarin	Barasingha
Lion-tailed macaque	Okapi
Gorilla	Gaur
Orangutan	Arabian oryx
Bonobo	Scimitar-horned oryx
Asian small-clawed otter	

The details of this program are documented in a master plan that provides institution-by-institution and animal-by-animal recommendations for management. Organizationally, each SSP is administered by a species coordinator assisted by a management committee elected by and from participating institutions.

Facilities for Captive Programs

Acceptance of captive propagation as an integral part of conservation strategies necessitates the consideration of optimal facilities. Two alternatives are apparent: (1) development of special facilities (those like Sybille or Patuxent that are extensions of the field programs may also require input from zoo professionals), or (2) use of existing zoos. Both have advantages and disadvantages.

Special facilities are able to concentrate on the species of concern. Moreover, such facilities can frequently be located near the natural habitat, which may be beneficial. Certainly, proximity to the natural habitat can facilitate movement between the wild and captive populations. One possible disadvantage is that development of such facilities and expertise may divert resources from other conservation activities. Development of a facility like Sybille is an expensive and arduous enterprise. Once numbers of black-footed ferrets permit, it will be imperative to distribute the captive population to more than one facility to avoid the risks of "all the eggs in one casket." Development of many special facilities like Sybille does not seem optimal, however, considering the limited resources available for wildlife conservation.

Utilization of zoos for captive propagation of black-footed ferrets can provide two advantages:

1. Zoos possess considerable resources and expertise that can be exploited for conservation purposes. Much of the funding for zoos is provided for reasons other than conservation. Zoos traditionally have evolved to serve local constituencies and communities. Even though zoos may now consider conservation their highest priority, the bases of support are still local; there is a great need to provide tangible gratification to these constituencies and communities in the form of the zoo. However, conservation has the opportunity to exploit the considerable resources of zoos by developing captive-propagation programs that an be significant and perhaps vital to the survival of species. This fact will be particularly relevant because of the need to distribute captive populations over multiple facilities.

2. Zoos can perform a significant educational service to conservation programs, especially for a species like the black-footed ferret. Many endangered species do not occur in the part of the planet where the majority of zoos are located. Consequently, the impact of educational efforts
may often be too remote or indirect to influence those who will really
decide the fate of wildlife. The black-footed ferret, however, is endemic
to the country with the largest number of zoos. Public attitude appears to
have been an important factor in the decline of the black-footed ferret.
Zoos could help to change these attitudes by encouraging more
awareness and appreciation of this species. Black-footed ferrets may not
be obviously "charismatic megavertebrates," but they are endearing
creatures that could be exhibited with prairie dogs, which are always
popular attractions.

Optimally, captive-propagation programs for black-footed ferrets
should encompass both special facilities like Sybille and traditional zoos
that demonstrate a commitment to and capability for propagation of the
species.

The success of captive propagation will depend, of course, upon mas-
tery of the basic reproductive husbandry of black-footed ferrets. Population management can be applied only when ferrets reproduce in numbers sufficient to generate a population.

Three final observations about the value of captive programs seem
appropriate. First, the biological value of captive propagation as reinforcement for wild populations has not been fully appreciated. Even
where its significance has been recognized, such a program is frequently
not implemented until wild populations are critically low. And there
may be a tendency to discontinue captive programs too soon after some
initial recovery of numbers in the wild. A population biology perspective
suggests, however, that even populations in the hundreds or thousands
may not insure long-term survival. In relation to black-footed ferrets, it
should therefore not be concluded that restored wild populations of
several hundreds, even if well protected, obviate the need or desirability
of a well-managed, back-up captive population.

Second, there has also been some criticism that captive propagation is
not a cost-effective way to conserve species. However, the temporary
untenability of a natural habitat may render captive propagation the
only way to preserve a species. If the conservation strategies exploit the
already existing resources for captive propagation, the cost-effective-
ness concern may not apply.

Third, previous lack of success in captive propagation of black-footed
ferrets (Carpenter 1985; Carpenter and Hillman 1978; Hillman and

Carpenter 1983) should be instructive but not dissuasive. Much progress has occurred in the last decade in the art and technology of captive propagation, especially in the areas of population management, basic husbandry, and veterinary care.

Population Management

As discussed elsewhere in this book, small populations, wild and captive, must be managed genetically and demographically if they are to survive as evolutionarily viable entities over the long term (Schonewald-Cox et al. 1983). The temporal implications of this point are important. It has been argued that the persistence, even expansion, of small populations over a period of a few years or perhaps decades invalidates concerns about genetic diversity or demographic stability. The recent history of the black-footed ferret in the wild should be sufficient evidence to refute such arguments. Moreover, arguments that cyclic behavior, even with wide amplitude, might have been normal for wild populations fail to appreciate that extinctions of local populations are probably common in species with population oscillations. When not only the numbers within local populations but also the number of such demes is low, and their isolation great, large oscillations in number are not viable.

There will be no attempt in this chapter to discuss extensively appropriate genetic and demographic management, but a few points seem pertinent. It should also be reemphasized that similar principles of population biology and management will apply to both small *in situ* and *ex situ* populations.

Genetically, small populations tend to lose variation important for species at both the population and individual level (Frankel and Soule 1981; Soule and Wilcox 1980). At the population level, genetic variation is required to permit continual adaptation to ever-changing environments. At the individual level, loss of genetic variation frequently produces inbreeding depression manifested by reduced survival and fecundity (Ballou and Ralls 1982; Soule 1980).

For very small populations, demographic stochasticity may cause greater risks than genetic problems (Goodman 1987; Samson et al. 1985; Shaffer and Samson 1985; Harris et al., chap. 6, this volume). Included among these risks are: devastation by natural catastrophes, decimation by disease epidemics, predator or competitor eruptions, and stochastic events in the survival or fertility of a small number of individuals (for example, all the offspring produced by the few adults in the population could be of the same sex).

Considering these genetic and demographic problems, an SSP will entail formulation of a master plan for populational management. The master plan must establish the genetic and demographic objectives for the program and then delineate the genetic and demographic management that will be applied to achieve these objectives.

Ultimately, of course, the genetic objective will presumably be for the forces of genetic change in populations to establish an equilibrium that will permit the species to continue to survive and evolve. However, there will be periods for many species, certainly the black-footed ferret, when population sizes necessary to attain such equilibria are impossible. Genetic objectives must then consider (Soule et al. 1986; Lande and Barrowclough 1986):

1. the kind (average heterozygosity versus allelic diversity) and level (90%, 95%) of genetic variation that must be maintained for the species to survive inside or outside its natural habitat. Preserving rarer alleles will require larger minimum viable populations (MVPs) than will merely maintaining average heterozygosity. Preserving 95% of average heterozygosity will require an MVP twice as large as 90% will;

2. the period over which these kinds and levels of diversity must be maintained (50 years or 200 years). In other words, how long will it be until some kind of viable genetic equilibrium can be established?

Attempts to determine an MVP must also consider problems of demographic and environmental stochasticity (Goodman 1987; Belovsky 1987; Shaffer 1987). As mentioned above, depending on conditions, these risks may be of comparable or greater significance than genetic problems to the population under consideration or management (Goodman 1987; Belovsky 1987). Analogous to the genetic considerations, demographic objectives that will influence MVP are: (1) the probability of persistence (50%, 90%, 95%) desired for the population; and (2) the period of time this probability of survival is required.

An important point to appreciate is that insuring 90% to 95% probability of persistence will require much larger populations than will average (50%) likelihood or survival. Statistically, because of the apparent pattern (negative exponential distribution) of extinctions due to these causes, 70% of actual populations would go extinct before the average (50%) time of persistence (Shaffer 1987). Thus, a 90% probability of survival for some given period of time would require the same size population as needed for average survival (50%) for 10 times as long; and 95% for 20 times as long (Belovsky 1987). In other words, in terms of population size required, 90% probability for 100 years is equivalent to 50% persistence for 1,000 years; 95% for 100 years is equivalent to 50% for 2,000 years.

These considerations will in general require that the SSP establish an MVP to achieve the objectives (Soule 1987a). This MVP will depend not only on the objectives suggested above, but also on certain biological characteristics of the population:

1. The generation time of the population (under existing or managed patterns of survivorship and fertility), which will vary with the value of demographic parameters that will be regulated by ecological conditions and/or by population management.

2. The N_e/N ratio in the population, which depends on the dynamics of propagation in the population (the sex ratio and family sizes) whether these are managed or not (Crow and Kimura 1970). Management can theoretically enlarge the N_e/N ratio.

3. The degree of subdivision or fragmentation in the population. Subdivision can be either a problem for or a benefit to genetic management depending on the relative sizes of the subpopulations and the total number in the overall population. Migration among subpopulations will often need to be managed to achieve some compromise between panmixia and subdivision. Depending on the amount of gene flow, natural or managed, subdivision may increase or decrease the N_e. Genetically, management of small and fragmented wild populations will normally require periodic movement of genetic material among the separate demes or subpopulations (Fig. 15.2). How much managed migration for genetic reasons will depend on whether or not the species or subspecies is to be managed as a single large population or several small populations (Shaffer and Samson 1985; Wilcox and Murphy 1985). There are trade-offs (Lacy and Clark, chap. 7, this volume). If the N_e of the combined subpopulations is much greater than the MVP, then it may be best

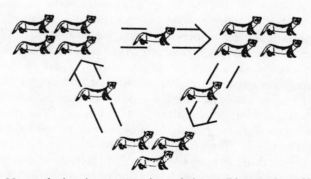

Figure 15.2. Managed migration among subpopulations (wild or captive) of black-footed ferrets.

to manage them as a single large interbreeding population. This strategy will require at least one migrant among each subpopulation per generation. However, if the N_e is less than the MVP for the intended objectives, it is possible to use the fragmentation of the population to advantage (Lande and Barrowclough 1987; Foose et al. 1986). Subdivision of small populations will actually increase the amount of allelic diversity preserved in the entire population. Each subdivision loses genetic diversity faster than a single large population of the same size as all the subdivisions combined, but different alleles are lost from the various demes, and overall allelic diversity is greater. To exploit this strategy, less than one migrant per population per generation is indicated.

Movement of animals among subpopulations may also be necessary for demographic reasons to reinforce or reestablish resident populations that have been decimated. Conversely, it may also be necessary to regulate the sizes of populations by removing animals to prevent overpopulation and consequent damage to the environment (Goodman 1980). Thus, demographic as well as genetic management will be in order.

4. The effective number of founders that have established the population or subpopulations, a most important consideration when determining the number of animals needed to found a viable captive population or to reestablish a population in a natural habitat.

5. The demographic rates of change in the population, especially during growth from foundation number to carrying capacity.

6. The kinds and intensity of genetic forces operating on the population, that is, is random drift dominant or are selection, mutation, or migration operating to mitigate its effects?

It is important to realize that there is no single magic number that represents the MVP for all species or even the MVP for any species all the time or under all circumstances (Soule 1987b). Rather, the MVP is established through consideration of the biological objectives and characteristics described above.

Once the MVP is established, the objective will be to stabilize the population or populations at some carrying capacity that is equal to or greater than the MVP, adjusted to allow for the N_e/N ratio. The genetic considerations will prescribe an MVP in terms of N_e. Normally, N_e is significantly less than N, however, due to the dynamics of propagation, natural or managed, in the population. Thus, to determine the actual number of animals (N) required to attain the MVP, the N_e generated by the genetic considerations will have to be divided by the N_e/N ratio. As other chapters discuss (Lacy and Clark, chap. 7, and Brussard and

Gilpin, chap. 4, this volume), accurate estimation of N_e is a formidable task.

Finally, populational analysis and management, particularly at the level being proposed for the captive population, will be possible only if adequate data on the vital statistics of individuals in the population are available. Thus, it is vital that good studbook-like data be maintained for the captive population and perhaps for wild populations during early stage of development.

Species like the black-footed ferret will survive only if conservation strategies are predicated on principles of population biology, incorporate both *in situ* and *ex situ* methods, and opportunistically exploit and coordinate all resources available for gene-pool preservation. Therefore, a survival plan employing the small remaining population of the species should be developed.

This species survival plan should:

1. Incorporate both captive and, when possible, wild populations as equally important and interactive components.
2. Utilize both special facilities like Sybille and qualified zoos for the captive propagation.
3. Establish MVPs that will include the total number of animals in the overall captive and wild population as well as the number and size of the subpopulations over which these animals are to be distributed.
4. Provide recommendations on genetic and demographic management: animal-by-animal and facility-by-facility for the captive population; at least subpopulation-by-subpopulation for the wild population.
   These recommendations should include: mate selection, reproduction schedules, and culling regimes (who should reproduce with whom, when, and how many times) in the captive population; and founder constitution and managed migration (who should be removed from or introduced into the subpopulations) in the wild.
5. Organize an effective management group to develop the SSP. The model provided by SSPs in the zoo community may be useful. This management group should include wildlife managers from states with wild populations and representatives of the various captive facilities participating in the propagation program for gene-pool preservation. It should also include external experts in population, reproduction, and other fields of biology.

References

American Association of Zoological Parks and Aquariums. 1983. *Species survival plan.* Wheeling, W.V.: AAZPA.

Ballou, J., and K. Ralls. 1982. Inbreeding and juvenile mortality in small populations of ungulates: A detailed analysis. *Biol. Cons.* 24:239–72.

Belovsky, G. 1987. Extinction models and mammalian persistence. In *Viable populations*, ed. M. E. Soule. Cambridge: Cambridge Univ. Press.

Carpenter, J. W. 1985. Captive breeding and management of black-footed ferrets. *Proceedings of the 1984 black-footed ferret workshop*. Laramie, Wyo.: Wyoming Game and Fish Dept.

Carpenter, J. W., and C. N. Hillman. 1978. Husbandry, reproduction, and veterinary care of captive ferrets. *Proceedings of the 1978 AAZPA annual conference*. Wheeling, W.V.: American Association of Zoological Parks and Aquariums.

Crawford, M. 1985a. The last days of the wild condor? *Science* 229:844–45.

———. 1985b. Condor agreement reached. *Science* 229:1248.

Crow, J. F., and M. Kimura. 1970. *An introduction to population genetics*. New York: Harper and Row.

Dresser, B. L., et al. 1986. Artificial insemination and embryo transfer in the Felidae. In *World conservation strategies for tigers*, ed. R. L. Tilson and U. S. Seal. Park Ridge, N.J.: Noyes.

Food and Agricultural Organization of the United Nations. 1986. *The Przewalski horse; and restoration to its natural habitat in Mongolia*. Rome: FAO.

Foose, T. J. 1983. The relevance of captive populations to the strategies for conservation of biotic diversity. In *Genetics and conservation*, ed. C. Schonewald-Cox, S. Chambers, B. MacBryde, and W. L. Thomas. Reading, Mass.: Addison-Wesley.

———. 1986. Species survival plans and overall management strategies. In *World conservation strategies for tigers*, ed. R. L. Tilson and U. S. Seal. Park Ridge, N.J.: Noyes.

Foose, T. J., and U. S. Seal. 1986. Species survival plans for large cats in North American zoos. In *Proceedings of the international cat congress*, ed. D. Miller. Washington, D.C.: Natl. Wildlf. Fed.

Foose, T. J., et al. 1985. Conserving animal genetic resources. *IUCN Bull.* 16(1–3):20–21.

Foose, T. J., et al. 1986. Propagation plans. *Zoo Biol.* 5(2):139–46.

Frankel, O. H., and M. E. Soule. 1981. *Conservation and evolution*, New York: Cambridge Univ. Press.

Goodman, D. 1980. Demographic intervention for closely managed populations. In *Conservation biology*, ed. M. Soule and B. Wilcox. Sunderland, Mass.: Sinauer Associates.

———. 1987. The demography of chance extinction. In *Viable populations*, ed. M. E. Soule. Cambridge: Cambridge Univ. Press.

Hillman, C. N., and J. W. Carpenter. 1983. Breeding biology and behavior of captive black-footed ferrets. *Intern. Zoo Yearbook* 23:186–91.

Kleiman, D., and B. Beck. 1986. Reintroduction of the golden lion tamarin to its natural habitat in Brazil. In *Primate conservation—the road to self-sustaining populations*, ed. K. Benirschke and D. Lindburg. New York: Springer-Verlag.

Lande, R., and G. Barrowclough. 1987. Effective population size, genetic variation, and their use in population management. In *Viable populations*, ed. M. E. Soule. Cambridge: Cambridge Univ. Press.

Samson, F. B., et al. 1985. On determining and managing minimum population size. *Wildl. Soc. Bull.* 13:424–33.

Schonewald-Cox, C., et al. 1983. *Genetics and conservation.* Reading, Mass.: Addison-Wesley.

Seal, U. S., and T. J. Foose. 1983a. Development of a masterplan for captive propagation of Siberian tigers in North American zoos. *Zoo Biol.* 2:241–44.

———. 1983b. Siberian tiger species survival plan: A strategy for survival. *J. Minn. Acad. Sci.* 49(3):3–9.

Seal, U. S., et al. 1986. Estrous cycles in Siberian tigers. In *Tigers of the world*, ed. R. L. Tilson and U. S. Seal. Park Ridge, N.J.: Noyes.

Shaffer, M. L. 1987. Minimum viable populations: Coping with uncertainty. In *Viable populations*, ed. M. E. Soule. Cambridge: Cambridge Univ. Press.

Shaffer, M. L., and F. B. Samson. 1985. Population size and extinction: A note on determining critical population sizes. *Am. Nat.* 125(1):144–52.

Smith, J. D. L., et al. 1986. Female land tenure systems in tigers. In *Tigers of the world*, ed. R. L. Tilson and U. S. Seal. Park Ridge, N.J.: Noyes.

Soule, M. E. 1980. Thresholds for survival: Maintaining fitness and evolutionary potential. In *Conservation biology*, ed. M. E. Soule and B. A. Wilcox. Sunderland, Mass.: Sinauer Associates.

Soule, M. E., ed. 1987a. *Viable populations.* Cambridge: Cambridge Univ. Press.

Soule, M. E. 1987b. Introduction and where do we go from here? In *Viable populations*, ed. M. E. Soule. Cambridge: Cambridge Univ. Press.

Soule, M. E., and B. A. Wilcox. 1980. *Conservation biology.* Sunderland, Mass.: Sinauer Associates.

Soule, M. E., et al. 1986. The millennium ark: How long the voyage, how many staterooms, how many passengers? *Zoo Biol.* 5(2):101–14.

Stanley-Price, M. 1986. Reintroduction of the Arabian oryx to its natural habitat in Oman. *Intern. Zoo Yearbook* 24.

Sullivan, A. L., and M. L. Shaffer. 1975. Biogeography of the megazoo. *Science* 189(4196):13–17.

Wilcox, B. A., and D. D. Murphy. 1985. Conservation strategy: The effects of fragmentation on extinction. *Am. Nat.* 125:879–87.

16

E. TOM THORNE AND DAVID W. BELITSKY

Captive Propagation and the Current Status of Free-Ranging Black-Footed Ferrets in Wyoming

Captive Propagation in Wyoming: By E. Tom Thorne

The black-footed ferret is probably the most endangered mammal on earth. The few ferrets remaining near Meeteetse and those captive at the Sybille Wildlife Research Unit of the Wyoming Game and Fish Department are the only representatives of this small, secretive weasel. The future existence of the species now appears to depend upon captive breeding.

In December 1981, the Wyoming Game and Fish Department (WGFD) was designated the lead agency in management of the black-footed ferret in Wyoming by the U.S. Fish and Wildlife Service (FWS) under authority of the Endangered Species Act of 1973. An interagency Black-Footed Ferret Advisory Team (BFAT) was formed to serve in an advisory capacity (Harju 1985). The team also included members representing private landowners with ferret habitat and nongovernmental conservation organizations. Following discussions on captive breeding of black-footed ferrets, BFAT and WGFD chose to take a conservative approach toward planning and initiation of captive propagation. The primary objective of the WGFD was and is to maintain a free-ranging population of black-footed ferrets if possible.

It took 2 to 3 years of research by teams fielded by FWS and Biota Research and Consulting, Inc., to determine 2 important points that WGFD considered relevant to captive propagation of black-footed ferrets. First, the ferrets apparently were limited to a complex of prairie dog towns west of Meeteetse with no other ferrets in the surrounding area (Fagerstone and Biggins, unpubl.; Groves and Clark 1986). The Meeteetse population is isolated and receives no genetic input by way of immigration. Second, the population was shown to be increasing to a peak of about 120 animals in 1984 (Fagerstone and Biggins, unpubl.; Wyoming Game and Fish Department, unpubl.).Thus ferrets probably were not yet suffering effects of inbreeding; or, at least, it was not being manifested by depressed reproduction (Ralls et al. 1979; Frankel and Soule 1981). In addition, the research suggested that the population was near its carrying capacity and that there was probably a small surplus of animals for captive breeding. Data generated in 1985 by recapturing ferrets ear-tagged in 1983 and 1984 demonstrated a very high rate of loss of juveniles (greater than 80%) by the time they reach 1.5 years of age (Fagerstone and Biggins, unpubl.; Wyoming Game and Fish Department, unpubl.). In 1984, 1 of 9 ferrets tagged as juveniles and 3 of 9 tagged as adults (more than 1 year old) in 1983 were recaptured; and in 1985, 0 of 49 animals tagged as juveniles and 5 of 25 as adults in 1984 were recovered. No juveniles or adults trapped in 1983 were retrapped in 1985. Some of these potentially surplus animals could be used as founder animals for captive breeding or to fill territories of adults removed for captive propagation.

A small isolated population, such as that at Meeteetse, exists in a precarious state and is vulnerable to extinction from random environmental events including loss of habitat, disappearance of prey, predation, and disease. A continuing loss of genetic diversity with the hazards of inbreeding depression can also occur. The research at Meeteetse, however, indicated that the black-footed ferret colony was thriving and in no obvious danger. Precautions to prevent human-related introduction of disease were initiated under direction of the WGFD when FWS and Biota began research in the summer of 1982 (Thorne et al. 1985).

Discovery of the Meeteetse population rekindled interest in ferrets and renewed hope that other colonies would be located. Studies at Meeteetse demonstrated that late summer and early fall, when the young begin venturing out of their natal burrows and dispersing from litter groups, are the best times to locate ferrets by spotlighting (Clark et al. 1983). Searching for tracks and characteristic ferret diggings in snow during winter is another, less efficient search technique (Clark et al. 1983; Clark et al. 1984). By late 1984, these techniques were well established, and failure to discover other ferret colonies using these and other methods was considered ominous. Thus, until another colony was located, the Meeteetse animals had to be managed as if they were the only surviving black-footed ferrets.

Captive propagation of black-footed ferrets in Wyoming under the leadership of the Game and Fish Department is appropriate. The department has a legislative mandate to serve as the main proponent for all species of wildlife in the state (Wyoming Statute 23-1-302). It was also felt that since the goal of captive propagation of ferrets would be re-introduction into their former habitat, they should be reared in habitat as nearly similar as possible to minimize the scope of acclimatization needed for the release program. Finally, there is the need for local governments, in this case the state of Wyoming, to conserve and manage their own populations of an endangered species. It has been estimated that in the near future as many as 2,000 species of terrestrial vertebrates will be in jeopardy of extinction unless protected, at least temporarily, by captive propagation (Soule et al. 1986). Obviously, this number is more than any single country, agency, or institution can accommodate. Some responsibility must therefore be shouldered by local governments. Because black-footed ferrets are regarded as a national treasure and because captive-reared ferrets will eventually be reintroduced into many or all of the states they formerly inhabited, cooperation and assistance of the federal government, other states, and nongovernmental conservation organizations are necessary.

The FWS attempted captive propagation of black-footed ferrets at the Patuxent Wildlife Research Center in the early 1970s (Carpenter 1985). Nine animals were taken into captivity from South Dakota. Four were killed by vaccine-induced canine distemper (Carpenter et al. 1976). Black-footed ferrets are so exquisitely susceptible to distemper that a vaccine found safe in domestic ferrets and Siberian polecats (apparently the nearest relatives to black-footed ferrets) produced a fatal disease rather than immunity. Two litters were produced from 1 female, but no offspring survived more than a few days. Poor reproductive success was suggested to be the result of inbreeding depression (Carpenter 1985; Carpenter and Hillman 1979; Hillman and Carpenter 1983).

Initially, WGFD and BFAT felt it was premature to begin captive breeding. The two groups felt the status and distribution of the Mee-teetse colony should first be determined and hoped that additional populations would be found. In the fall of 1983, a WGFD proposal to consider captive propagation (Thorne, unpubl. mimeo) was rejected as premature. In April 1984, a meeting to discuss captive breeding was held in Cheyenne at WGFD headquarters. It was attended by representatives from FWS, the New York Zoological Society and Wildlife Preservation Trust International; Biota; the University of Wyoming and Washington State University; and several state wildlife management agencies. Although there was no consensus on captive propagation and no indication that funding for captive propagation was available, an ad hoc Captive Breeding Committee representing FWS, universities, nongovernmental organizations, and WGFD was appointed to consider requirements of ferret captive breeding.

The committee presented its report and recommendations (Anderson et al., unpubl. mimeo) to BFAT and WGFD that summer. The report described the basic facilities and personnel necessary for captive propagation and made the following recommendations:

1. A captive breeding program involves a cooperative effort between federal, state and private conservation agencies. The U.S. Fish and Wildlife Service must be a willing participant and make a commitment to provide funds and serve as coordinator.
2. A black-footed ferret research/breeding review team should be established as soon as possible to consider captive breeding as a recovery alternative for black-footed ferrets. If captive breeding is considered a viable alternative, recommendations should be made for possible facility construction in 1985.
3. Every effort should be made to obtain funds. Once base funding is available, a foundation (e.g., Peregrine Fund) might be established to assist with the cost of operations. The source of funding might influence site location.
4. A site selection committee should consider visiting potential breeding sites.
5. Consideration should be given to removing young from the Meeteetse population during the fall of 1984 and keeping them in a temporary facility.

In September 1984, BFAT voted that captive propagation should be undertaken in Wyoming and recommended a site analysis of several locations near Laramie. The advisory team agreed there should be a technical committee to advise on captive breeding and voted against taking ferrets into captivity in 1984 due to lack of both funding and a permanent captive-breeding facility. These recommendations were provided to WGFD. Also in September 1984, a black-footed ferret workshop in Laramie was sponsored by the Wyoming Cooperative Fisheries and Wildlife Research Unit, FWS, and WGFD. This workshop compiled most of the knowledge and research results generated up to that time by the Meeteetse site field investigations (Anderson and Inkley 1985). The program included a panel on captive propagation and there was much discussion of captive breeding during the workshop. One paper recommended early capture of a few ferrets (Bogan 1985).

In September 1984, the Game and Fish Department submitted a proposal to the FWS for cooperation and funding for captive propagation to be done in the Laramie area. In addition, the Wyoming Game and Fish Commission voted in October to "go on record as wanting to keep the ferrets in Wyoming and establish a research station (captive-breeding program) in Wyoming and authorize the Department's personnel, based on all information available, to decide where in Wyoming to locate the research station, and that the Department be empowered and authorized to apply to the Endangered Species Fund and to do what is necessary to save the ferret in Wyoming." A site analysis (Reese, unpubl. mimeo) was submitted to FWS in January 1985. The site analysis was conducted from a property-right perspective pursuant to the bundle of rights theory whereby each right represents a distinct and separate right or privilege of ownership; the most suitable property is the one with the greatest number of rights. Of the 7 potential sites analyzed, the Sybille Wildlife Research Unit of the WGFD was recommended as the most suitable location for development of a black-footed ferret captive-propagation center.

In April 1985, both the chief engineer and wildlife veterinarian of WGFD visited the FWS Patuxent Wildlife Research Center in Laurel Md.; the Bronx Zoo of the New York Zoological Society in New York City; and the Wildlife Conservation and Research Center of the Smithsonian Institution in Front Royal, Va. The purpose of these visits was to gain additional insights into the physical requirements for captive propagation of a small carnivore.

In May 1985, the University of Wyoming pledged support to the WGFD for a captive-propagation effort if it was deemed necessary for recovery of the black-footed ferret. Also in May 1985, a ferret recovery meeting was held at the FWS regional office in Denver. It was attended by representatives of many states formerly inhabited by ferrets. During that meeting, FWS representatives were optimistic that federal funds for captive breeding would be available in 1986 or 1987. Consequently, a decision was made to capture ferrets in the fall of 1985 after population estimates were completed. Captive animals were to be a hedge against a catastrophe in the wild and to serve as founder animals for captive breeding. Six to 10 animals were to be captured yearly from 1985 to 1987. This would provide adequate founder animals without seriously reducing the free-ranging population. Ferrets captured in 1985 were to be held in temporary quarters at the veterinary laboratory at the Sybille unit.

During that meeting, it was also agreed that, under the leadership of the FWS, a multistate committee would be formed to develop an objective technique for evaluating and prioritizing future reintroduction sites. In addition, a proposal for black-footed ferret captive breeding prepared by Washington State University and submitted to the FWS by Wildlife Preservation Trust International was considered. The proposal was rejected, in part because of previous positions of BFAT and the Wyoming Game and Fish Commission that ferrets should not be removed from Wyoming.

During the summer of 1985, a second captive-propagation proposal was prepared by WGFD and submitted to the FWS in September. Modifications were made at the existing Sybille veterinary laboratory to provide for protection of captive ferrets from disturbance and diseases, both animal and human in origin. Plans were made for capture and transport of 6 ferrets in late September and early October. Ferrets were to be taken 2 at a time with a 10- to 14-day interval between captures to assure that the animals were healthy, adjusting to their cages, and accepting food. Three females and 3 males were to be captured in order to provide optimum genetic diversity in founder animals. Because it was felt that successful breeding would require mature, experienced males, 1 or 2 mature males were to be captured. Even so, emphasis the first year was to be on acclimatizing them to captivity rather than on breeding.

In late September and early October 1985, 6 ferrets were captured near Meeteetse and moved to Sybille. The second 2 were captured 23 days after the first 2, which had remained healthy and adjusted to captivity. The last 2 were obtained 1 and 6 days later as field biologists completed the last of the 1985 population estimates. Two mature males were included with the 6 ferrets, and the intended accumulation of select founder animals seemed to have an outstanding start.

Optimism was soon replaced by dismay. The 1985 population estimates showed the Meeteetse ferret population had declined from approximately 58 in August to an estimated 31 animals in October (Fagerstone and Biggins, unpubl.). The FWS and WGFD called a meeting in Cheyenne on October 22, 1985, to discuss the decrease and possible courses of action. It was attended by field biologists and several authorities on the biology and demographics of small populations.

The day before that meeting, the last captured ferret died of a disease characterized by thickening and reddening of the skin and intense itching. The other ferret captured at the same time had the same illness. It was apparent both had contracted the disease while in the wild. Canine distemper was tentatively diagnosed the night of October 21, 1985; at the meeting the next morning, the disease and its consequences were announced. Because all captive ferrets had come from the same colony and presumably were exposed to the same diseases, they had not been isolated from each other when taken to Sybille. Our emphasis was to protect them from diseases coming from outside the colony rather than from within. Therefore, it was predicted that the other 4 captive ferrets would die, which they did. Diagnosis of canine distemper in the Meeteetse ferret colony explained why ferret numbers were declining; the epizootic would probably kill the remaining free-ranging ferrets.

After considering recommendations from the meeting, it was decided to attempt to save a remnant of the population. Past experience had shown that capturing every ferret would be very time-consuming, if not impossible. Capture locations for the 2 animals already diseased when trapped indicated that the distemper was distributed across the central portion of the colony. Although research had failed to reveal ferrets in small, remote prairie dog towns surrounding the central colony, failure in 1985 to recapture any ferrets ear-tagged as juveniles in 1984 suggested that they had dispersed from the main colony and that a few might be surviving in peripheral prairie dog towns. Consequently, a decision was made to trap all ferrets that could be located in the core colony and to leave any animals existing in outlying areas. The rationale was that ferrets from the core area were probably doomed to die from distemper, but that some could be trapped before being exposed to distemper and used for captive breeding. By removing ferrets from the core colony, it was hoped that their exposure to the disease in remote towns would be greatly lowered and that some would escape exposure and survive. Survivors might repopulate the core prairie dog towns after canine distemper had run its course.

Every ferret located (2 juvenile males and 4 females) was captured. They did not develop distemper. As in the case of ferrets that died, these animals were vaccinated with an inactivated canine distemper virus vaccine prepared by Dr. M. J. G. Appel of the James A. Baker Institute for Animal Health at Cornell University, except that an experimental adjuvant provided by Fort Dodge Laboratories was used. Two ferrets were tested by serum neutralization and shown to have protective antibodies.

After the quarantine period and thorough disinfection of the ferret quarters with sodium hypochlorite, these 6 ferrets were moved to Sybille. Despite intensive efforts to encourage breeding, the 2 juvenile males were behaviorly and physiologically immature and infertile. Hopes for captive breeding had to wait until 1987.

Partial first-year funding for captive breeding was approved by the Wyoming Game and Fish Commission in January 1986 and by the FWS in February 1986. Allocations for the project totaled $285,795 in 1986–87, $234,827 of which came from the federal grant. A building containing 4 large rooms for ferret cages was constructed at Sybille. In addition, it included an office-commissary, a shower-in complex for visitors, a medical treatment room, and isolation rooms. A biologist-veterinarian was assigned full time to the project.

Late in 1985, the recommendation of the ad hoc Captive Breeding Committee regarding a technical advisory committee was fulfilled when the Captive Breeding Specialist Group (CBSG) of the Species Survival Commission, International Union for the Conservation of Nature and Natural Resources (IUCN), was invited to fill that role. That highly

respected group has the backing of the IUCN and other nongovernmental conservation organizations. The CBSG can call on authorities in
a variety of fields related to captive breeding for assistance. Site reviews
have been completed, and each has generated specific recommendations to help assure success in ferret captive propagation.

As of March 1987, there were 18 black-footed ferrets in captivity.
Three were parous, mature females at least 4 years of age, 2 were parous
females at least 1 year of age, 1 was a nulliparous female 1 year old, and 5
were juvenile females. There was 1 mature male that may have never
participated in breeding activities and 1 mature male that may have
sired the 2 litters present in 1986, two 1-year-old males that did not
breed in 1986, and 3 juvenile males. All captive ferrets were doing well.

Status of Free-Ranging Black-Footed Ferrets at Meeteetse in 1986:
By David W. Belitsky

Censuses of the black-footed ferret colony in northwestern Wyoming
have been conducted since its discovery in September 1981. The study
area, located 15 km west of Meeteetse, Wyoming, includes over 33
white-tailed prairie dog (*Cynomys leucurus*) towns totaling more than
2,886 hectares (Forrest et al. 1985a). This section summarizes results of
surveys conducted during the periods November 11, 1985, to March 19,
1986, and July 13 to August 12, 1986. Surveys have been completed in
all areas occupied by ferrets in previous years. Surveys in 1986 included
the objective of transferring some of the free-ranging ferrets to the
captive population.

The winter and summer surveys followed the guidelines set forth in
Clark et al. (1984). Additional winter surveys using aircraft were conducted by personnel of the U.S. Fish and Wildlife Service and the Wyoming Game and Fish Department, as suggested by Hendersen et al.
(1969). The 1986 study area (Fig. 16.1) included prairie dog towns that
either had not been intensively surveyed in the past because they lacked
a history of black-footed ferret occupation or were not located until that
year (Forrest et al. 1985a, 1985b; Richardson et al. 1984; Clark et al.
1983). A comparison of the current study area with that discussed in
Forrest et al. (1985a,12) shows an additional 600 hectares surveyed in
1986.

    I would like to recognize the assistance of personnel of the University of Wyoming
Cooperative Wildlife Research Unit and its leader, Dr. Stanley Anderson, the U.S. Fish
and Wildlife Service, and Brian Miller and Jon Hanna of the Wyoming Game and Fish
Department in this survey effort.

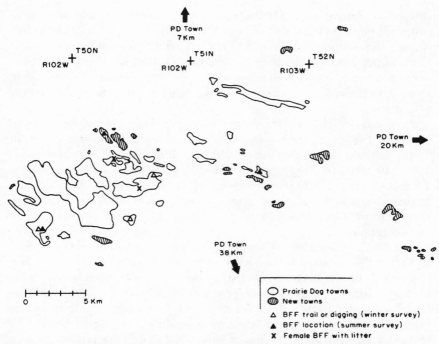

Figure 16.1. Distribution of prairie dog towns in the complex at Meeteetse, Wyoming, as of August 1986, showing locations of black-footed ferret (BFF) trails from the winter survey, locations of animals from the 1986 summer survey, and locations of females with litters.

The evaluation of the 1986 summer survey includes the following account of cumulative effort. (Searches—known as spotlighting—were conducted using high-intensity automobile-battery-powered spotlights, which detect the eye shine of the ferret.) One survey night was tallied for each night that one or more observers spent spotlighting in the field; and a spotlight hour was tallied for each hour that an observer spent searching either from a vehicle or on foot with a backpack. Through August 11, an average of 10 observers were fielded on each of 30 survey nights, expending 1,100 spotlight hours, compared to 544 and 647 spotlight hours during 1984 and 1985 respectively.

During 6 days of the winter survey, 3 black-footed ferret trails or diggings were located on 4 prairie dog towns (Fig. 16.1). Persistent snow cover suitable for recording ferret trails is necessary for thorough winter surveys. During the 1985–86 winter, however, snow cover persisted only during November 10–15, 1985.

Winter survey techniques produce results that should be interpreted cautiously. Biggins et al. (1986) reported on activity patterns of a radio-collared ferret that was not active above ground for 5 consecutive days during the winter of 1981–82. This behavior emphasizes the importance of persistent snow cover for thorough winter surveys. In previous winters, at least 3 periods of persistent snow cover occurred, resulting in more complete surveys and more accurate estimates of the winter population.

Winter surveys of the Meeteetse complex from December 12, 1984, to March 5, 1985, resulted in observation of 37 ferret trails or diggings in 14 prairie dog towns. For the period December 20, 1983, to March 31, 1984, 49 trails or diggings were observed on 17 prairie dog towns. Therefore, the 1985–86 winter survey results yielding 3 trails on 4 prairie dog towns indicated a much lower population level than recorded in previous years, even considering the limited opportunity to record ferret activity.

As a result of the summer survey, 2 adult females with litters of 4 and 5 as well as 3 individual ferrets were observed in 5 different prairie dog towns (Fig. 16.1). One of the single ferrets was known to be an adult male and was added to the captive population on July 27. Two juvenile females removed from the litter of 4 and 1 juvenile female from the litter of 5 were added to the captive population on August 11 and 12 respectively. The juvenile females from the 2 litters differed by as much as 2 weeks in maturity (Thorne, telephone conversation, 1986).

On August 27 and 28, 1986, WGFD and FWS considered the current status of black-footed ferrets at the Meeteetse site and, with reluctance, agreed to implement a WGFD contingency plan for capture of ferrets in 1986. Because of the small number of litters present, the recommendation to capture as many ferrets as possible was enacted. Two mature females and 5 juveniles were captured. Consequently, a total of 11 black-footed ferrets were captured for captive propagation in 1986. The last known ferret, an adult male, was captured in February 1987 and added to the captive colony.

References

Anderson, S. H., and D. B. Inkley, eds. 1985. *Black-footed ferret workshop proceedings.* Cheyenne, Wyo.: Wyoming Game and Fish Dept.

Ball, S., and D. Belitsky. 1985. Completion report, plague control, Meeteetse, Wyoming, insecticide dust application, July–September, 1985. Cheyenne, Wyo.: Wyoming Game and Fish Dept. Mimeo.

Biggins, D. E., et al. 1986. Activity of radio-tagged black-footed ferrets. *Great Basin Nat. Memoirs* 8:135–40. Provo, Utah: Brigham Young Univ.

Bogan, M. A. 1985. Needs and directions for future black-footed ferret research. In *Black-footed ferret workshop proceedings*, ed. S. H. Anderson and D. B. Inkley, 28.1–28.5. Cheyenne, Wyo.: Wyoming Game and Fish Dept.

Carpenter, J. W. 1985. Captive breeding and management of black-footed ferrets. In *Black-footed ferret workshop proceedings*, ed. S. H. Anderson and D. B. Inkley, 12.1–12.13. Cheyenne, Wyo.: Wyoming Game and Fish Dept.

Carpenter, J. W., and C. N. Hillman. 1979. Husbandry, reproduction, and veterinary care of captive ferrets. 1978 *Proc. Amer. Assoc. Zoo Veterinarians* 36–47.

Carpenter, J. W., et al. 1976. Fatal vaccine-induced canine distemper virus infection in black-footed ferrets. *J. Amer. Vet. Med. Assoc.* 169(9):961–64.

Clark, T. W., et al. 1983. Handbook of methods for locating black-footed ferrets. *Wyoming Bur. Land Mgmt. Wildl. Tech. Bull.* no. 1. Cheyenne, Wyo.: U.S. Bur. Land Mgmt. and Wyoming Game and Fish Commission.

Clark, T. W., et al. 1983. Report to Wyoming Game and Fish Commission. Black-footed ferret advisory team litter survey and minimum population count of BFF at Meeteetse. Wyoming Game and Fish Dept. Mimeo.

Clark, T. W., et al. 1984. Handbook of methods for locating black-footed ferrets. *Wyoming Bur. Land Mgmt. Wildl. Tech. Bull.* no. 1. Cheyenne, Wyo.: U.S. Bur. Land Mgmt.

Clark, T. W., et al. 1984. Seasonality of black-footed ferret diggings and prairie dog plugging. *J. Wildlife Management* 48:1441–44.

Dexter, W. D. 1985. Introductory remarks. In *Black-footed ferret workshop proceedings*, ed. S. H. Anderson and D. B. Inkley, 1.1–1.2. Cheyenne, Wyo.: Wyoming Game and Fish Dept.

Fagerstone, K. A., and D. E. Biggins. 1986. Summary of black-footed ferret and related research conducted by the Denver Wildlife Research Center, 1981–1986. Denver Wildlife Research Center, U.S. Fish and Wildlife Service.

Forrest, S. C., et al. 1985a. Final report, black-footed ferret population status at Meeteetse, Wyo. Wyoming Game and Fish Dept. Mimeo.

Forrest, S. C., et al. 1985b. Black-footed ferret habitat: Some management and reintroduction considerations. *Wyoming Bur. Land Mgmt. Wildl. Tech. Bull.* no. 2. Cheyenne, Wyo.: U.S. Bur. Land Mgmt.

Frankel, O. H., and M. E. Soule. 1981. *Conservation and evolution*. New York: Cambridge Univ. Press.

Franklin, I. R. 1980. Evolutionary change in small populations. In *Conservation biology: An evolutionary-ecological perspective*, ed. M. Soule and B. Wilcox, 135–49. Sunderland, Mass.: Sinauer Assoc.

Groves, C. R., and T. W. Clark. 1986. Determining minimum population size for recovery of the black-footed ferret. *Great Basin Nat. Memoirs* 8:150–59.

Harju, H. J. 1985. Black-footed ferret advisory team efforts. In *Black-footed ferret workshop proceedings*, ed. S. H. Anderson and D. B. Inkley, 4.1–4.4. Cheyenne, Wyo.: Wyoming Game and Fish Dept.

Henderson, F. R., et al. 1969. The black-footed ferret in South Dakota. *S. Dak. Dept. Game, Fish and Parks Tech. Bull.* no. 4. Rapid City, S. Dak.: S. Dak. Dept. Game, Fish and Parks.

Hillman, C. N., and J. W. Carpenter. 1983. Breeding biology and behavior of captive black-footed ferrets. *Intern. Zoo Yearbook* 23:186–91.

Madson, C. 1985. Ferrets need your help. *Wyoming Wildlife* 49(11):10–13.

May, R. M. 1986. The cautionary tale of the black-footed ferret. *Nature* 320:13–14.

Ralls, K., et al. 1979. Inbreeding and juvenile mortality in small populations of ungulates. *Science* 206:1101–03.

Richardson, L., et al. 1984. Progress report: Winter black-footed ferret conservation studies. Cheyenne, Wyo.: Wyoming Game and Fish Dept. Mimeo.

Soule, M., et al. 1986. The millennium ark: How long a voyage, how many staterooms, how many passengers? *Zoo Biol.* 5:101–13.

Thorne, E. T., et al. 1985. Capture, immobilization, and care of black-footed ferrets for research. In *Black-footed ferret workshop proceedings,* ed. S. H. Anderson and D. B. Inkley, 9.1–9.8. Cheyenne, Wyo.: Wyoming Game and Fish Dept.

17

MICHAEL W. DONCARLOS, BRIAN MILLER, AND E. TOM THORNE

The 1986 Black-Footed Ferret Captive-Breeding Program

Most species in the family Mustelidae have been difficult to breed in captivity. Of 28 species inventoried by the International Species Information System (ISIS) during the past 5 years, only 2, ermine (*Mustela erminea*) and spot-necked otters (*Lutra maculicollis*), have had a captive birthrate that exceeded the death rate (ISIS 1985). For these two species, the captive birthrates at individual institutions (the Minnesota and Brookfield Zoos respectively) were responsible for the total captive population growth; if these institutions are not taken into account, however, then all 28 mustelid species inventoried by ISIS failed to increase in population size during the past 5 years. Thirteen of 28 species (46%) declined in population size. The assumption that breeding is a priority may not be valid for all captive mustelid species. Striped skunks (*Mephitis mephitis*) and domestic ferrets (*Mustela putorius*) are maintained but not bred by many ISIS participants. Comparison of captive-born (cb) versus wild-caught (wc) recruits into the captive population may be a better indicator of captive-breeding success in individual species. In the past 5 years, both cb and wc animals were recruited into the captive populations of 21 mustelid species; only 10 of those species recruited more animals through captive breeding than through wild capture.

Several factors may be responsible for the poor captive-breeding performance within the Mustelidae. Many captive mustelid species are maintained at low populations. Although mean population size was 22, those populations in which the cb/wc recruitment ratio was less than 0.7 averaged only 7 individuals, while populations in which the cb/wc ratio was greater than or equal to 0.7 averaged 49 individuals (ISIS 1985). Many potential wild-caught founders failed to reproduce, presumably due to failure to adapt to captive conditions. Of particular significance to the black-footed ferret project is the high captive-reproductive failure rate of males reported in a number of mustelid species, including 3 of 4 congeneric species (DonCarlos et al. 1986; Enders 1952; Heidt et al. 1968; Liers 1951; Wright 1948; and Wuestenberg, pers. comm.).

The reproductive biology of the black-footed ferret is incompletely understood and has been reported for only 4 captive animals (2 males and 2 females) monitored in 4 breeding seasons during 1975–78. One female copulated 3 out of 4 years and whelped 2 litters. The other female failed to copulate with either male (Carpenter 1985). Hillman and Carpenter (1983) concluded that the black-footed ferret was seasonally monoestrous and reported that females entered proestrus from February 23 to March 15. Copulation was prolonged, and female precopulatory behaviors were not remarkable. Forrest et al. (unpubl.) reported 1 female black-footed ferret reaching sexual maturity at 1 year of age. Atherton (pers. comm.) reported that 2 male black-footed ferrets were aspermic when examined at about 1 year of age.

The 1986 black-footed ferret reproductive management strategy was jointly developed by J. Doherty, M. DonCarlos, U. Seal, and R. Wuestenberg, members of the International Union for Conservation of Nature and Natural Resources/Species Survival Commission/Captive Breeding Specialist Group and by T. Thorne and B. Miller, staff members of the Wyoming Game and Fish Department. The strategy considered previous captive black-footed ferret results (Carpenter 1985) and was modeled after proven captive-breeding methods developed for wild-caught ermine (DonCarlos et al. 1986) and mink (Enders 1952).

Methods

Six black-footed ferrets captured during October and November 1985 at Meeteetse were managed for natural breeding in 1986. Ages, sex, and other individual animal data are in Table 17.1.

We assumed that social behavior and spacing patterns in black-footed ferrets were similar to the general pattern in *Mustela*, in which large male territories overlap 1 or more smaller female territories, and both sexes defend their territories against same-sex intruders (Powell 1979). Radiotelemetry observations of spacing patterns of black-footed ferrets at Meeteetse were consistent with this assumption (Biggins et al. 1985). Consequently, animals were housed in individual cages. Cage arrangement was linear; male 655 and female 659 had constant visual access to each other, but all other animals were in visual isolation.

The ferrets were housed in cages constructed of wood and vinyl-clad welded wire that measured 81 cm wide, 152 cm long, and 61 cm high. Adjacent cages had shift doors for introducing or transferring animals.

Table 17.1. Description of Individual Black-Footed Ferrets in Captivity

Identification No.	Sex	Age	Capture	Comments
655	M	Juvenile	10/31/85	Half sibling (same sire) of 657
657	M	Juvenile	10/25/85	Son of 659, half sibling (sire) of 655
407	F	Adult	10/25/85	Parous (in wild)
577	F	Adult	10/29/85	Parous (in wild)
659	F	Adult	11/02/85	Parous (in wild), mother of 657
688	F	Juvenile	10/31/85	None

Each cage had 1 or more nest boxes, which were used for animal transfers during breeding introductions.

All ferrets were housed in the same room. Ambient temperature in the room was maintained between 7° and 18° C. Photoperiod was natural; windows provided most of the light, and fluorescent lights were controlled by timers within the incident photoperiod.

The ferrets were fed whole rodent carcasses—mice (*Mus musculus*), hamsters (*Mesocricetus auratus*) and prairie dogs (*Cynomys* sp.)—6 days a week, and canned cat food the remaining day. Food was provided once daily; and the amounts presented and consumed were recorded. Fresh drinking water was provided *ad libitum*.

Breeding Management and Behavior

Breeding introductions began on February 17, 1986, and continued through April 27, 1986. Introductions ($N = 99$) occurred at all hours, but most were made between 0900 and 1700 hours. Durations varied from several minutes to several hours; most were between 1½ to 2 hours. Females were transferred to the males' cages either by shift doors or by nest box transfer. Because all animals were distracted by the presence of human observers, breeding introductions were monitored from a remote location with Panasonic WV 1850 TV cameras equipped with Fujinon CF 12.5 A-SND-P 1 : 1.4/12.5 lenses, Panasonic WV 5410 monitors, and a Panasonic AG-6010 VCR. Breeding introductions were recorded on VHS videocassettes and cataloged for later analysis.

Following copulation with male 655, female 659 was transferred to an isolation room furnished with 3 nest boxes. Lactating domestic ferrets were made available as mother surrogates if natural rearing or lactation failed.

A general description of black-footed ferret breeding introductions is provided in Table 17.2. All interactions were initiated by the females. The 2 males differed in their reaction to the presence of females. Male 657 reacted in a defensively aggressive manner to all females introduced to him, regardless of their behavior. By contrast, male 655 was passive or indifferent to most females introduced to him. Because of this difference, male 655 was given greater exposure to females.

Differences in the behavior of the 4 females were also apparent. The yearling female 688 reacted in a defensively aggressive manner to both males, and her behavior did not change or moderate with repeated introductions. The adult female 407 remained passive and rarely approached either male. Adult females 577 and 659 approached the males or ignored them, but seldom displayed defensive or aggressive behavior. Four pairings between 655 and 659 (from February 18 to 21, 1986) and 3 pairings between 655 and 577 (from February 24 to 27, 1986) resulted in solicitous female behavior. Male 655 made no attempts to mount or restrain female 577 in response to her solicitations. However, 655 made several unsuccessful mounting attempts in response to solicitations of 659 on February 18, 1986, and copulated with 659 on February 20 and 21, 1986.

Chuckling vocalizations were heard during several breeding introductions, including the 2 copulatory bouts. Vocalizations were not monitored systematically.

Six different male breeding behaviors were observed and are described:

1. MGS (Male Genital/anal Sniff)—The male sniffs the female genital or anal area.
2. M (Mount)—The male places his front legs over the female's back and grasps her firmly. His body is aligned directly over hers with his head directly above her head, neck, or upper back.
3. MNL (Male Neck Lick)—The male licks the female's neck in short rapid motions.
4. MNG (Male Neck Grasp)—The male grasps the female by the nape of the neck. The bites may be short and quick or may take the form of a single, long grasp. Occasionally, the male may jerk the female's head and neck in a backward motion or drag her entire body backwards.
5. PT (Pelvic Thrust)—The male moves his hips in a thrusting motion while mounted over the female.
6. LS (Leg Swim)—During intromission the hind legs of the male are raised off the floor. The hind legs then move in a circular motion consisting of raising the leg, moving it forward in an extension, pushing it down, and drawing it back to the original position.

Table 17.2. Black-Footed Ferret Pairings from February 17, 1986, through April 27, 1986

Male × Female	No. of Pairings[a]	No. of Copulations	No. of Female Solicitations
657 × 688	3	0	0
657 × 577	4	0	0
657 × 407	14	0	0
657 × 659	3	0	0
655 × 688	9	0	0
655 × 577	31	0	3
655 × 407	17	0	0
655 × 659	18	2	17

[a]$N = 99$.

Eight different female breeding behaviors were observed and described:

1. FA (Female Approach)—The female approaches the male in a low crouch with head and neck extended. She moves toward the male in the typical quadruped cross-step fashion and keeps her eyes fixed on him. There may be pauses in the approach, but her eyes do not stray. Her tail is still and usually extended.

2. HUC (Head Under Chin)—The female crouches and places the top of her head under the male's chin. She may hold that position for varying lengths of time.

3. FRS (Female Rubs Side)—The female extends her head toward the male and touches her nose and cheek to the neck or cheek area of the male. She then rubs her head down the length of his body while arching her body toward the male so that her entire side rubs against him. She may also rub her head forward toward the male's nose while arching her body sideways against him. She may then rub the entire side of her body against his cheek as she moves forward. A third variation involves the female moving under the male's chin as she arches her body sideways against him. When the middle of her back is under his chin, she may raise her back in an arching motion to lift the male's head and neck as she passes under him. All of these rubs expose the female genital area to the male's head. When the rub is complete, she may circle back and pause next to the male.

4. HUM (Head Under Midsection)—The female approaches the male in a crouched position (usually from an angle) and pushes her head under his midsection. She may then nudge him or simply hold her head stationary.

5. CU (Crawl Under)—After the female pushes her head under the male, she may push down and forward to crawl under the male's body. She may either pause directly under the male or pass entirely underneath his body.

6. FNB (Female Neck Bite)—The female grips the male's neck with her teeth.

7. FGS (Female Genital Sniff)—The female sniffs the male's genital or anal area.

8. HS (Head Sway)—During copulation, the female holds her head low and swings it from side to side. The head may be twisted slightly so that the bottom of the jaw is at an angle to the floor. The head sway is seen when the male is gripping the female's neck with his teeth.

Three introductions in which female 577 solicited male 655 were similar. All 3 sequences began with a female forward approach (FA). During the first 2 of these introductions, a female rub against the side of the male (FRS) followed the FA. In the first introduction, several FRS episodes occurred. Male 655 responded to 1 FRS with a genital sniff (MGS), but showed no further interest, and the female retired to a corner. In the second introduction, female 577 pushed her head under the male's midsection (HUM) after the FRS and sustained this position for the next 16 minutes. Male 655 remained motionless and otherwise unresponsive. In the third introduction, female 577 followed the FA by placing her head under the male's chin (HUC). Six seconds later she moved away, paused momentarily, and bumped her head against the male's cheek. In a continuous motion she then arched her body sideways against his (FRS), pushed her head under his chin, and arched her back as it passed under his chin, thus raising his head and neck. The male remained motionless and otherwise unresponsive, and the female then retreated.

By contrast, female 659 was more persistent in her solicitation of male 655, approaching the male 17 times. In 11 of these solicitations, male 655 failed to respond. The sequence of behaviors in these introductions is described in Table 17.3.

On 4 occasions, solicitous behavior by female 659 was followed by mounting (M) by male 655 but without intromission. Three of these mounts took place on a day in which no copulation occurred (February 18, 1986), and one mount preceded the successful mounting and copulation that occurred on February 20, 1986. During these noncopulatory mounts, the male spent about 75% of the time licking the female's neck (MNL) or grasping her neck with his teeth (MGN). The sequences of female solicitation behaviors preceding these mounts are described in Table 17.4.

Table 17.3. Sequence of Female Solicitations When Male Remained Unresponsive in Introductions between Male 655 and Female 659

	Introduction										
	1	2	3	4	5	6	7	8	9	10	11
	FA	FA	FA	FA	FA	FA	FA	FA	FA	FA	FNB
	HUC	FNB	FRS	FRS	FRS	FRS	HUC	HUC	HUC	CU	HUM
	FRS	FRS	HUC	FGS	FNB		CU	CU	HUM	HUM	
		HUC		HUM	FRS		FGS	CU	FRS		
		HUC			HUM				HUM		
		FRS			CU						

Note: See text for explanation of abbreviations.

Table 17.4. Sequence of Female
Solicitations Preceding Male Mounting
in Introductions between Male 655
and Female 659

	Introduction		
1	2	3	4
FA	FA	FA	FA
FRS	FRS	HUC	HUM
HUM	FNB	FRS	CU
		HUC	
		FRS	
		CU	

Note: See text for explanation of abbreviations.

Male 655 and female 659 copulated for 17 minutes and 24 seconds
on February 20, 1986. Intromission was interrupted twice by the
female during this copulatory bout, and the copulation was preceded
by 2 mountings that did not result in copulation. The sequence and
duration of behaviors in this copulation are described in Table 17.5.

This same pair copulated again on February 21, 1986, for 7 minutes
and 3 seconds, with 1 disruption of intromission. The sequence and
duration of behaviors in this copulation are described in Table 17.6.
Pregnancy did not result from copulations between male 655 and
female 659.

The copulatory behaviors we observed in black-footed ferrets were
similar in several details to descriptions of copulatory behaviors previously reported for the European ferret. Male 655 spent about 75% of
his copulatory time either grasping (MGN) or licking (MNL) the
female's neck. The male European ferret also either grasps or licks the
female's neck during copulation (Murr 1931). In both observed black-
footed ferret copulations, intromission was initiated by a series of pelvic
thrusts, comparable to those reported for the European ferret (Poole
1966). Pelvic thrusts were only occasionally observed after initial penetration, and there were periods during copulation in which little or no
motion occurred. These periods of immobility were similar to those
previously reported for the black-footed ferret (Hillman and Carpenter 1983) and the European ferret (Murr 1931).

We believe that intromission occurred during both black-footed ferret copulations. Three specific behaviors—female head sway (HS),
male leg swim (LS), and a lateral orientation of the female's tail—are
considered diagnostic of intromission in ermine (DonCarlos, pers. observ.), and mink (Wuestenberg, pers. comm.). We observed all of these

Table 17.5. Sequence and Duration
(in Minutes and Seconds) of Behaviors
during a Copulation between Male 655
and Female 659 on February 20, 1986

Female		Male	
FA	:28		
FRS	:01		
HUC	:03		
FNB	:01		
CU	:10		
		M	:05
		Dismount	
		MNG	:04
		M	:01
		MNG	:10
		Dismount	
		M	17:24
		PT	:02
HS	:03	MNG	:11
		MLN	:03
		MNG	:16
HS	:03	MNG	6:17
		PT	:03
		LS	:04
HS	:11	LS	:14
		LS	:14
HS	:10	LS	:17
		LS	:05
HS	:04	LS	:02
HS	:09	LS	:18
		LS	:16
Coitus interrupted, still mounted			
		MNG	:08
		MNL	:09
HS	:04	MNG	4:09
		PT	:09
HS	:03	LS	:20
		LS	:05
Coitus interrupted, still mounted			
		MNG	3:03
HS	:06	PT	:08
		LS	:11
		Dismount	

Note: See text for explanation of abbreviations.

Table 17.6. Sequence and Duration
(in Minutes and Seconds) of Behaviors
during a Copulation between Male 655
and Female 659 on February 21, 1986

Female		Male	
FA	:06		
FRS	:02		
		MNG	:06
		M	7:03
		MNG	3:25
HS	:03	PT	:08
HS	:06		
Coitus interrupted, still mounted			
		PT	:10
		LS	:18
		MNL	:03
		MNL	:14
		MNG	:25
		LS	:05
Coitus interrupted, still mounted			
		MNG	1:53
		PT	:15
		LS	:05
		LS	:09
HS	:23	LS	:23
HS	:25		
		Dismount	

Note: See text for explanation of abbreviations.

behaviors in black-footed ferrets but do not know if ejaculation occurred. Observations of copulations in 1987 and 1988 suggest that
these behaviors in black-footed ferrets are not always diagnostic of intromission. In retrospect, we are unsure if intromission occurred during the 1986 copulatory bouts.

We observed 20 female solicitations. The behavioral sequences were
similar, regardless of differences in male responses. Detailed reports
describing copulatory behavior in the European ferret do not describe
female solicitations preceding male mounting (Poole 1966, 1974; Murr
1931). By contrast, female solicitous behavior preceding copulation has
been observed in ermine (DonCarlos, pers. observ.) and mink
(Wuestenberg, pers. comm.) when the male involved is passive or inex-
perienced. It is not observed when an aggressive male is involved.

Chuckling vocalizations were heard during several black-footed fer-

ret introductions and were similar to precopulatory vocalizations of ermine (DonCarlos, pers. observ.) and mink (Wuestenberg, pers.
comm.). Hillman and Carpenter (1983) reported low hissing and
whimpering by a female black-footed ferret during copulation, but did
not describe these vocalizations in comparison with other mustelid species. Precopulatory chuckling or clucking vocalizations have been reported in ermine (DonCarlos et al. 1986), long-tailed weasels, *Mustela
frenata* (Wright 1948), mink (Enders 1952), and the European ferret
(Poole 1966).

Three of the 4 female ferrets in this study were known to have reproduced previously in the wild, and 2 of these 3 actively solicited
males during the 1986 breeding season. We therefore conclude that
the failure to breed more than 1 female and to produce litters was due
to male reproductive failure.

Males 655 and 657 were remarkably different behaviorally. Male 655
was passive in response to female solicitation, but did copulate following persistent solicitation from female 659. By contrast, male 657 was
defensively aggressive in response to all female approaches, including
solicitation by female 659 on the same day that she copulated with male
655. Similar variance in male breeding behavior has been reported in
ermine (DonCarlos et al. 1986), mink (Enders 1952), river otters, *Lutra
canadensis* (Liers 1951), and long-tailed weasels (Wright 1948). Male reproductive variance can preclude successful breeding when population
size is small.

The 2 different male behavioral patterns described here are identical
to patterns of behavior observed in individual ermine (DonCarlos,
pers. observ.) and mink (Wuestenberg, pers. comm.). In addition, electroejaculation attempts of males 655 and 657 in April 1986 failed to
yield sperm, and the testes of both males were small (Atherton, pers.
comm.). Enders (1952) reported that not all male mink reach sexual
maturity at 10 months of age. Both males may have been sexually immature during the 1986 breeding season. Consequently, if the behavior and development of these 2 black-footed ferrets were comparable
to that of ermine and mink, we predicted that male 655 would reach
behavioral and physiological reproductive maturity and breed successfully in the 1987 breeding season. By contrast, we felt that male 657
was unlikely to ever become a successful breeder. Thereafter, male 655
sired one litter in 1987 and 5 litters in 1988; male 657 failed to copulate
with any females in 1987 and 1988.

We conclude that supervised, direct introduction and behavioral
testing of potential breeders, independent of or in addition to physical
estrus indicators, are the best methods of captive-breeding manage-

ment for many mustelid species, including black-footed ferrets. This methodology has the following advantages: (1) it is proven for several mustelid species; (2) it is effective in the absence of detailed knowledge of breeding seasons and physical estrus indicators; (3) it ensures that copulation is observed; and (4) it allows multiple breeding combinations and so facilitates clear evaluation of individual animal reproductive performance.

References

Biggins, D. E., et al. 1985. Movements and habitat relationships of radio-tagged black-footed ferrets. In *Black-footed ferret workshop proceedings.* Cheyenne, Wyo.: Wyoming Game and Fish Dept.

Carpenter, J. W. 1985. Captive breeding and management of black-footed ferrets. In *Black-footed ferret workshop proceedings,* Cheyenne, Wyo.: Wyoming Game and Fish Dept.

DonCarlos, M. W., et al. 1986. Captive biology of an asocial Mustelid: *Mustela erminea. Zoo Biol.* 5:4.

Enders, R. K. 1952. Reproduction in the Mink (*Mustela vison*). *Proc. Am. Phil. Soc.* 96:6.

Heidt, G. A., et al. 1968. Mating behavior and development of least weasels (*Mustela nivalis*) in captivity. *J. Mamm.* 49:3.

Hillman, C. N., and J. W. Carpenter. 1983. Breeding biology and behaviour of captive black-footed ferrets. *Intern. Zoo Yearbook* 23.

International Species Inventory System (ISIS). 1985. Species Distribution Report for 31 December 1985, pp. 2095–269.

Liers, E. E. 1951. Notes on the river otter (*Lutra canadensis*). *J. Mamm.* 32:1.

Murr, E. 1931. Observations on the mating of the ferret. *Zool. Garten,* Leipzig, N.F. 4:289–91.

Poole, T. B. 1966. Aspects of aggressive behavior in polecats. *Zeitschrift für Tierpsychologie* 24:351–69.

———. 1974. The effects of oestrus condition and familiarity on the sexual behavior of polecats (*Mustela putorius* and *M. furo* × *M. putorius* crosses). *J. Zool.* 172:357–62.

Powell, R. A. 1979. Mustelid spacing patterns: Variation on a theme by Mustela. *Zeitschrift für Tierpsychologie* 50:153–65.

Wright, P. W. 1948. Breeding habits of captive long-tailed weasels (*Mustela frenata*). *Am. Midland Nat.* 39:2.

18

JONATHAN D. BALLOU AND ROBERT OAKLEAF

Demographic and Genetic Captive-Breeding Recommendations for Black-Footed Ferrets

In August 1986, the Captive Breeding Specialist Group of the IUCN met in Laramie, Wyoming, with personnel of the Wyoming Game and Fish Department and the U.S. Fish and Wildlife Service to discuss a management plan for the captive population of black-footed ferrets at Sybille, Wyoming. The plan was to address the genetic and demographic management of the captive animals as a population and to take as a priority the long-term preservation of the black-footed ferret as a species. This chapter details the recommendations that came out of that meeting, one of which was a propagation plan detailing:

a. Explicit goals for production in the F_1 and F_2 generations to achieve a secure, self-sustaining captive population as the precursor to a release and reintroduction program;

b. Explicit population criteria for use of black-footed ferrets from the Sybille colony to establish a second geographically separate breeding population. (These criteria were to be part of the above analysis.) It was suggested that if 3 litters (9–10 animals) were produced in the 1987 season, then a breeding nucleus could be selected and a second population established in time to prepare for their breeding in the 1988 season. In no way was this to jeopardize reproductive success at Sybille in 1988.

c. Explicit criteria for conserving the status of the captive population for the initiation of a release and reintroduction program. It was noted that the captive population will probably be about 200 breeding adults and will probably be F_2 or F_3 generation animals. Their removal was not to compromise the status of the captive population. It was thought that conducting small-scale reintroductions with carefully chosen animals would be appropriate in order to perfect techniques. Meeteetse was suggested as the most suitable site.

In addition, this chapter includes:

    d. Explicit recommendations for establishing future captive-population subdivisions of black-footed ferrets; and

    e. Recommendations on general strategies for maintaining genetic diversity in the captive population.

Goals and Priorities of the Ferret Breeding Program

In developing population management recommendations, it is imperative to recognize the primary goals and priorities of the captive-breeding program. They are as follows:

    a. The primary goal is the long-term preservation of the black-footed ferret as a species;

    b. The recent history of declining population trends and of disease susceptibility of the wild population suggests that long-term preservation necessitates the establishment of a secure, self-sustaining captive population;

    c. Reintroduction of black-footed ferrets back into available natural wild habitat should be done as soon as possible, while at the same time minimizing survival risks to the captive population;

    d. Maintenance of genetic diversity is of primary importance to the survival of both the captive and future wild populations.

Although the history of the Meeteetse population suggests that the population may have been isolated since the 1930s and could therefore have lost most of its genetic diversity (Lacy and Clark, chap. 7, this volume), prudent management dictates that measures be taken to preserve whatever diversity has been retained in the founding animals (Hedrick et al. 1986; Allendorf and Leary 1986). Therefore, once secure and over the critical phase, the captive population will be managed with the goal of maintaining genetic diversity.

Recommendations for a General Breeding Strategy

In light of these priorities and long-term goals, the following general strategies are recommended:

    a. The highest priority during the early critical phase of the captive population is to increase the population size as rapidly as possible. Rapid growth of the small founding population is imperative for both genetic and demographic reasons. Small population size increases the risk of extinction due to demographic stochastic events and reduces the retention of genetic diversity (Gilpin and Soule 1986). A rapid increase in the population size mitigates both these risks.

b. When establishing pairings, those that maximize the retention of founder alleles in the offspring should be considered first. This must be done, however, without compromising population growth. In selecting mates, genetic compatibility should be secondary to reproductive compatibility. Genetic compatibility should consider minimizing the levels of genetic relatedness between mates and assuring a diverse representation of founder alleles in future generations (Foose 1983). This is initially accomplished by guaranteeing that a maximum number of offspring from each founder pair reproduce in subsequent generations. At least 7 offspring per founder pair are required to be 95% certain that all the founders' allelic diversity is passed on to the next generation (Ballou 1984).

c. Establishment of a breeding nucleus at a second site is recommended as soon as possible to mitigate the possibility of catastrophy or disease epidemics destroying the Sybille population (Dobson and May 1986).

d. Data should be collected throughout the development of the population and analyzed periodically to evaluate the fitness as well as genetic and demographic characteristics of the population. Types of data to be collected should include studbook data (AAZPA and IZY Studbook Procedures) as well as morphological and genetic data (Wayne et al. 1986).

e. Levels of genetic variation in existing ferrets should be evaluated through biochemical analysis. Such analyses (electrophoresis, mitochondrial DNA, and hypervariable DNA probes) may be useful in determining more accurately the relationships of the existing ferrets as well as paternity in cases where multiple matings occur. The goal to maintain genetic diversity and avoid breeding close relatives should not be abandoned if little or no genetic variation is found. Lack of variation does not imply total lack of genetic diversity, and it may be vital to maintain what little genetic variation is present for the long-term fitness of the population.

f. Individuals with unusual (outlying) reproductive or phenotypic fitness characteristics should not be culled or otherwise limited in reproductive potential. Neither the genetic variation nor demographic potential in these individuals should be prejudiced by the occurrence of potentially deleterious characteristics assumed to be genetic (Frankham et al. 1986).

g. When the initial critical phase of population expansion is complete (once 20 potential breeders are established in each of 2 populations; see below), more emphasis should be placed on maintaining genetic diversity. Such considerations include the following:

- Establish pairings to maximize the number of surviving founder alleles and compensate for earlier nonoptimal genetic pairings. This is accomplished through constant pedigree analysis and mate selection (Mac-Cleur et al. 1986).

- Utilize the genetic contributions of the original founders as long as possible, even if invasive techniques are required. This is accomplished through preferentially breeding earlier-generation as opposed to later-generation individuals (that is, by continuing to breed founders in preference to their offspring; F_1 in preference to F_2 and so on).

- Evaluate the effects of inbreeding on the fitness of the population utilizing data collected during the critical phase of population growth (Ballou, chap. 5, this volume).

- Expand the distribution and number of individuals into a number of population subdivisions. Continue to recommend high rates of migration between subdivisions until founder alleles are equally distributed among subdivisions. Once this is accomplished, limit the degree of migration within the total population to 1 effective breeder per generation, thereby establishing true genetic subdivisions, maximizing the retention of genetic diversity, and minimizing the potential for epidemiological disaster (Lacy 1987). Migration between subdivisions should be rotated among subdivisions. At least 5 population subdivisions are recommended for the black-footed ferrets (Lacy 1987).

- Establish an overall captive-population carrying capacity sufficient to retain a specified amount of the genetic diversity originally acquired by the founding animals. Initial demographic models suggest that 500 animals (200 breeding adults during the birth season) will retain 80% of the heterozygosity of the founders for 200 years (see Genetic Model below) (Soule et al. 1986).

- Through demographic management, maintain a stable captive population at or above this carrying capacity. Reproductive rates should be managed to support a stationary, stable population while supplying animals for reintroduction programs (Foose 1982; Goodman 1980).

h. Demographic and genetic analyses should be conducted routinely during the development of the population. Estimates of potential growth rates, generation length, effective population size, and rates of loss of heterozygosity should be updated. If necessary, the recommendations specified in this plan should be revised to adjust for more accurate estimates of these genetic and demographic parameters.

i. Genetic and demographic recommendations for the reintroduction program need to be developed. These should include discussions of 2-way exchange between captive and wild populations and of the selection of individuals for reintroduction.

Captive-Population Management Objectives and Recommendations

The specific management recommendations for the captive black-footed ferret population are based on the above strategies and were developed to achieve a series of sequential objectives. The basic objectives and their estimated time frames are summarized in Fig. 18.1. The recommendations for achieving the objectives are listed under each objective. The time frame for the program has not been emphasized since the future growth of the population is uncertain and specific time-oriented recommendations are not possible. However, the models do enable us to estimate potential time frames for the various stages of the management plan, and we have included the estimated schedule from

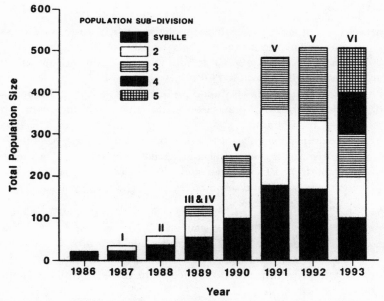

Figure 18.1. Estimated growth and distribution of the captive population of black-footed ferrets. Growth and time schedule estimates are based on the realistic demographic model. Roman numerals indicate the program objectives (see text).

the results of the realistic population model (see Demographic Models).

For purposes of the discussion below, the following terms are clarified:

• All ferrets brought in from Meeteetse are labeled "wild-caught." However, these wild-caught animals consist of both "founders" (by definition those ferrets that are unrelated to each other and whose genetic contributions are present in the living population) and F_1 generation animals (offspring of founders). The pedigree for the population with founders indicated as such is shown in Fig. 18.2.

• "Potential" breeders are individuals of breeding age whose behavior and reproductive characteristics do not preclude them from breeding.

Objective Level I (Estimated Year 1987): To produce enough young from the original 18 animals at Sybille to establish an F_1 and F_1 founder backcross breeding nucleus at a second population subdivision (POP2) geographically separated from Sybille.

The second population subdivision should be established under the following minimum constraints. If the constraints cannot be met, a second population should not be established and all the Sybille animals should be kept at Sybille until reproduction is such that the conditions are met:

    a. Selection of pairings at Sybille should be made in accordance with the recommendations for breeding specified above.

    b. A maximum of 10 animals, including a minimum of 3 males and 2 females, should be sent to POP2.

    c. The wild-caught animals should preferentially be kept at Sybille.

    d. Individuals for POP2 must be selected so that no pairings would occur between full siblings.

    e. In the event of highly successful reproduction during the first year, those offspring exceeding the initial needs of POP2 should be used first to establish 10.10 (10 males and 10 females) potential and proven breeders at Sybille and second to establish additional pairs at POP2.

Objective Level II (Estimated Year 1988): To build the population subdivisions at both Sybille and POP2 to at least 10.10 potential F_1, F_1/founder backcross, and founder breeders. (This objective may have already been accomplished in Objective I if reproduction at Sybille was outstanding.)

The following recommendations specify the order in which this objective should be accomplished:

    a. Breed animals in Sybille (founders in preference to F_1 or F_1 backcrossed to founder) to establish a breeding core of at least 10.10 potential breeders at Sybille.

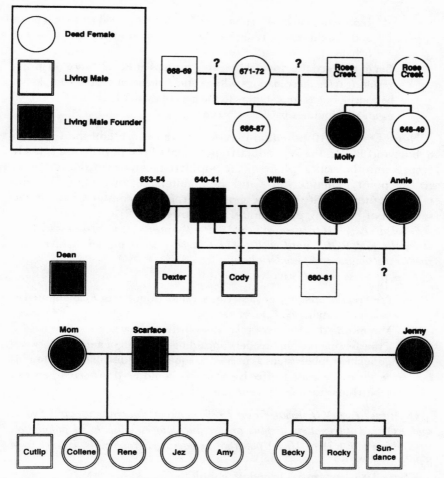

Figure 18.2. The pedigree of the 18 wild-caught black-footed ferrets at Sybille. Circles are females; squares are males. Living animals are shown with double squares or circles. Animals that are considered founders have been shaded. The Rose Creek male and female have never been in captivity. Animal 686–87 may have been sired by either the Rose Creek male or male 668–69. In addition, 1 of 3 females (Willa, Emma, or Annie) may have been the mother of Cody.

b. Send additional captive-bred F_1 and F_1/founder backcrosses to POP2 until POP2 has 10.10 potential F_1 breeders. Animals selected for POP2 should meet the following criteria:

   (1) Pairing combinations among animals sent to POP2 should not include full-sibling pairings;

   (2) The 10.10 potential breeders at POP2 should consist of a genetic

representation of all founder ferrets that have bred up to that point and should closely resemble the distribution of founder alleles at Sybille;

c. Begin to breed the original breeding nucleus at POP2. Since offspring resulting from these matings will be in the F_2 generation, they should not be considered as contributors to the goal of establishing 10.10 potential F_1 and F_1/founder backcross breeders.

Note: Once 10.10 potential breeders have been established at both Sybille and POP2, the reproductive potential of the population is fairly secure, and the critical phase of the population can be considered over. At this point, genetic criteria and the maintenance of genetic diversity take a higher priority and the general strategy recommended for the selection of breeding pairs should be followed.

Objective Level III (Estimated Year 1989): To maintain and breed individuals at Sybille and POP2 until each population has 10.10 proven breeders with similar distributions of founder alleles.

This process should involve:

a. Preferential breeding of earlier-generation (founder or F_1 as opposed to later-generation (F_2, F_3) ferrets;
b. Minimizing the degree of relatedness between mates;
c. Transferring proven breeders (including wild-caught animals) between population subdivisions as needed to equalize the distribution of founder alleles. At least 1 effective breeder should be transferred between subdivisions each ferret generation (2.5 years).

Objective Level IV (Estimated Year 1989): To reproduce individuals to Sybille and POP2 until enough surplus are produced to establish a 3d population subdivision (POP3) of 10.10 potential breeders geographically distinct from Sybille and POP2.

These 10.10 potential breeders should:

a. Be early generation (F_1 or F_2) ferrets;
b. Come from Sybille or POP2 as needed but not reduce either subdivision below 10.10 proven breeders;
c. Be representative of the overall founder allele distributions existing in Sybille and POP2;
d. Consist of potential pairings with degrees of relatedness no larger, and preferably lower, than the average degree of relatedness of breeding animals in Sybille and POP2.

Objective Level V (Estimated Year 1991–1992): To breed individuals within Sybille, POP2, and POP3 as 1 genetic population until a total effective population size of 250 is reached (estimated 500 animals, 200 breeding-aged adults during the birth season; see Genetic Model below). The carrying capacity will be considered 500 animals. (These carrying-capacity estimates are open to revision as additional information from the captive population is collected.)

a. Animals should be transferred between population subdivisions as needed to continue to equalize the distribution of founder alleles. At least 1 effective breeder should be transferred between population subdivisions per ferret generation (2.5 years);

b. Using existing data, demographic models should be updated to determine when reproduction will allow animals to be released;

c. A limited number of animals can be used for experimental reintroductions when it can be shown through demographic modeling using data from the existing population that Objective Level V can be obtained within 2 years following the experimental release.

Objective Level VI (Estimated Year 1992): To initiate a full-scale reintroduction program using animals that are surplus to the stable maintenance needs of the population; and, once the distribution of founder alleles across Sybille, POP2, and POP3 is equalized, to subdivide the total population into 5 genetically separate divisions with an effective size of 50 each.

a. Demographic modeling, based on data collected from the captive population to this point, should be initiated to determine reproductive rates of animals in Sybille, POP2, and POP3 necessary to maintain a stationary, stable captive population at carrying capacity (500 animals) and to supply the needs of a full-scale reintroduction program.

b. Individuals selected for the reintroduction program should be selected so that the level of genetic diversity in the captive population is not compromised (reduced) by their removal. Therefore, those individuals should be descended from founders whose genetic representation in the captive population is not affected by the individuals' removal. Pedigree analysis should be used to assist in determining individuals to be reintroduced.

c. Distribution of founder alleles among population subdivisions should be equalized through the transfer of animals.

d. The total captive population should be subdivided into 5 genetically separate (but not necessarily geographically separate) divisions with an effective size of 50 each.

e. To maximize the retention of genetic diversity, there should be limited gene flow between subdivisions. Only 1 effective migrant per generation (2.5 year) within the total population should be transferred between subdivisions.

f. Within each subdivision, pairings should be selected to minimize the degree of relatedness and to maximize the allelic diversity.
g. To assure continued epidemiological protection, the 5 subdivisions should be distributed over at least 3 geographically separate facilities.

Demographic Models

The purpose of developing a model of the growth and population characteristics of the captive colony of black-footed ferrets is twofold:

1. To establish a likely time frame for the initiation of different phases of the management program. For these estimates, population growth rates are required.
2. To evaluate the levels of genetic variation likely to be retained in the captive population after a specified amount of time. For these calculations, population growth rates, age structures, and generation lengths are required.

A deterministic model was used to simulate potential growth rates, generation lengths, age structures, and reproductive rates for a variety of population growth scenarios. A deterministic (as opposed to stochastic) model was chosen because a wide range of population outcomes could be simulated. Furthermore, reliable data on important life-history parameters in captivity were scarce if not nonexistent.

The population parameters used to model these scenarios were based on a variety of sources and assumptions. Unfortunately, the data required for developing accurate models are severely lacking. The most reliable information based directly on black-footed ferret data relates to the size of wild litters that survive until weaning. All other population parameters must be estimated from experience with other species in captivity.

Three scenarios were modeled. A "pessimistic" model depicts population trends if successful reproduction occurs but at a rate considered low relative to the reproductive potential of the species. (In a sense, the label is misleading since a truly pessimistic model would depict no reproduction followed by imminent population extinction.) A "realistic" model depicts the response of the population to reasonable reproductive and survival rates, while the "optimistic" model measures the response to relatively high survival and reproductive rates. Listed below are the estimates used in the various models, their sources, and assumptions.

There are no data documenting litter size at birth. However, counts at Meeteetse during weaning (July and August) indicate a maximum litter size of 5 and a mean litter size of 3.4 (Thorne, chap. 16, this volume). Counts during 1985 were not included in calculations of mean litter sizes due to the catastrophic events during that year. We assumed that litter sizes at weaning could be improved in captivity but used data from the wild as a conservative estimate in the model calculations. Assuming a litter size of 5 at birth, the survival rate from birth to weaning was 68%. This rate was used for the proportion of animals living to approximately 3 months of age.

Modeling of the Meeteetse population at carrying capacity suggests juvenile survival rates of 15% and 28% for males and females respectively (Harris et al., chap. 6, this volume). We assumed differential survival to be the result of territorial pressures in the wild and that in captivity the sexes would exhibit similar survival rates. We also assumed that much of the mortality experienced by juveniles in the wild could be avoided in captivity. Based on experience with captive ermine (*Mustela erminea*), we assumed that postweaning survival could range between 95% in the realistic case and 80% in the pessimistic case (DonCarlos, interview, 1986). As specified above, we assumed the survival rate from birth to weaning as 68%. Therefore, total first year survival was assumed to range between 54 and 65% for the various scenarios. Data from the International Species Information System (ISIS) on *Mustela putorius furo* and *Mustela erminea bangsi* born in captivity since 1980 show first-year survival rates of 73 and 82% respectively. Possible bias in these data (failure to report occurrence of early death) suggests that these rates may be overestimated. They nevertheless do provide estimates on upper ranges of expected survival rates. In addition, survival rates of 95% for domestic mink are obtainable and routinely observed (Johansson 1961).

It is difficult to determine adult survival of the Meeteetse population with available data. Modeling suggests, however, that adult survival may vary from 50 to 80% (Harris et al., chap. 6, this volume; Oakleaf, unpubl. data). Again, we assumed the survival would be greater in captivity and similar among sexes. Data on adult survival of other mustelids in captivity are not available. Yearly adult survival (px) in captivity was assumed to be relatively high, ranging from 80% for the pessimistic case to 95% for the more realistic case.

It was assumed that the sex ratio was 50 : 50 at birth and that there were no differences between the sexes in mortality rates.

One yearling female was known to produce young in the wild (Forrest et al. 1988). However, no data are available to suggest the percentage of yearlings that reproduce in the wild. Data from other mustelids, especially the American martin (*Martes americana*), influenced Harris et al. (chap. 6, this volume) in modeling of black-footed ferret populations to use 85% as the rate at which yearling females breed. In our modeling, we assumed no yearling production in the pessimistic scenario; 50% of the young reaching weaning were assumed to reproduce in the realistic case, and 80% in the optimistic case.

Data from the wild population show that at least 80% of the wild adult females were reproductively active and probably as many as 90 to 95% of the females in fact did produce young each year. Adult reproductive rates were assumed to be 50% for the most pessimistic case and 80% for the realistic and optimistic cases. Reproductive longevity was assumed to be 5 years, although it is likely that black-footed ferrets can breed beyond that age.

Longevity was assumed to be 10 years. Survival decreased linearly between the ages of 5 and 10 years at which point no animals remained alive.

The parameters used for the demographic simulations as well as the resulting demographic estimates are shown in Table 18.1. The pessimistic and optimistic simulations were derived by directly applying these parameters to the 1986 age distribution of the captive black-footed ferret. In the realistic simulation, however, a lag or delay in reproduction was incorporated into the model by forcing the model to produce only 10 young in 1987 and having only 50% of all reproductive-age animals

Table 18.1. Parameters and Results of the Demographic Model of Captive Black-Footed Ferrets

	Model		
Variable	Pessimistic	Realistic[a]	Optimistic
Litter size	5	5	5
Preweaning survival	68%	68%	68%
Weaning litter size	3.4	3.4	3.4
Postweaning to 1st-year survival	80%	95%	95%
Total 1st-year survival	54%	65%	65%
Yearlings reproducing	0%	50%	80%
2–5 year-olds reproducing	50%	80%	80%
Results:			
Generation length	2.6 yrs.	2.5 yrs.	2.4 yrs.
Growth per year	128%	202%	230%
Breeders[b]	23%	49%	60%

[a]During the first year of the realistic model, only 10 young were produced and allowed to survive to 1 year; during the second year, only 50% of the yearlings and adults bred. After the second year, the model used the values listed in the table.
[b]Percentage of animals that breed based on the stable-age distribution at the birth season during the growth phase (see Table 18.5).

reproduce the second year. This reduced production during the first and second breeding seasons was thought to be a likely result of establishing successful propagation techniques. The demographic parameters were then applied to the 1988 population to estimate future population characteristics.

Note: Demographic and genetic analyses should be conducted routinely during the development of the population, and estimates of potential growth rates, generation length, effective population size, and rates of loss of heterozygosity should be updated. If necessary, the recommendations specified in this plan should be revised to adjust for more accurate estimates of these genetic and demographic parameters.

Data from the simulated models, giving 9-year population projections and stable-age structures, are shown in Tables 18.2 through 18.4.

Genetic Model

The results from the demographic models were used to estimate potential effective population sizes for captive black-footed ferrets as well as the rate at which the captive population will lose heterozygosity. Both of these genetic parameters are important for determining the desirable carrying capacity of the captive population.

The measure of effective population size is useful for evaluating how well the population can retain long-term genetic diversity. Of primary interest is the ratio of effective size to census size (N_e/N). High ratios are indicative of population structures likely to retain genetic diversity, while low ratios indicate high rates of diversity loss.

The effective size of a population is determined by a number of variables: sex ratio of breeders, size of the breeding population, and the variance in the number of young produced (family size). Effective size is maximized by an equal sex ratio, a large number of breeders, and a low variance in family size—that is, each breeding unit producing the same number of young (Kimura and Crow 1963; Lacy and Clark, chap. 7, this volume). In captivity, management practices can manipulate breeding schedules so that the effects of unequal breeding sex ratio and variance in family size are somewhat mitigated. The proportion of individuals participating in breeding remains critical however.

Table 18.5 includes the percentage of the stable-age distribution of breeding animals for the 3 demographic models used. In the realistic and optimistic models, 50 and 80% of the yearlings reproduced respectively. The total proportion of the population that breeds during the year is therefore 23%, 49%, and 60%, respectively for the 3 models.

Table 18.2. Nine-Year Projections for the Pessimistic Demographic Model of Female Black-Footed Ferrets

Age Class	1986	1987	1988	1989	1990	1991	1992	1993	1994	1995
0	5	8	11	13	17	21	28	35	45	58
1	1	4	4	6	7	9	12	15	19	25
2	2	1	3	3	5	6	7	9	12	15
3	2	2	1	3	3	4	5	6	7	10
4	1	2	1	1	2	2	3	4	5	6
5	0	1	1	1	0	2	2	2	3	4
6	0	0	1	1	1	0	1	1	2	2
7	0	0	0	1	1	1	0	1	1	2
8	0	0	0	0	0	1	1	0	1	1
9	0	0	0	0	0	0	0	0	0	1
10	0	0	0	0	0	0	0	0	0	0
Total females	11	16	22	28	36	45	59	75	96	123
Total population	18	33	44	56	72	91	117	149	192	245

Note: Yearly growth rate (λ) = 1.28; generation time = 2.6 years.

Table 18.3. Nine-Year Projections for the Realistic Demographic Model of Female Black-Footed Ferrets

Age Class	1986	1987	1988	1989	1990	1991	1992	1993	1994	1995
0	5	5	12	40	74	150	304	611	1,240	2,516
1	1	4	5	8	26	48	97	196	395	801
2	2	1	5	5	7	25	46	92	186	375
3	2	2	1	5	5	7	23	43	87	177
4	1	2	2	1	5	4	7	22	41	83
5	0	1	2	2	1	4	4	6	21	39
6	0	0	1	2	2	1	4	3	5	18
7	0	0	0	1	1	1	1	3	3	4
8	0	0	0	0	1	0	1	0	2	2
9	0	0	0	0	0	0	1	1	0	1
10	0	0	0	0	0	0	0	0	0	0
Total females	11	15	28	63	121	241	486	979	1,982	4,017
Total population	18	27	56	126	242	483	973	1,958	3,963	8,034

Note: Yearly growth rate (λ) = 2.02; generation time = 2.5 years.

Table 18.4. Nine-Year Projections for the Optimistic Demographic Model of Female Black-Footed Ferrets

Age Class	1986	1987	1988	1989	1990	1991	1992	1993	1994	1995
0	5	19	43	99	224	514	1,187	2,737	6,303	14,517
1	1	3	12	28	64	145	332	767	1,768	4,072
2	2	1	3	12	27	60	138	316	728	1,680
3	2	2	1	3	11	25	57	131	300	692
4	1	2	2	1	3	10	24	55	124	285
5	0	1	2	2	1	3	10	23	52	118
6	0	0	1	2	1	1	2	8	19	43
7	0	0	0	1	1	1	1	2	7	15
8	0	0	0	0	0	1	1	0	1	5
9	0	0	0	0	0	0	1	1	0	1
10	0	0	0	0	0	0	0	0	0	0
Total females	11	28	64	146	332	761	1,753	4,039	9,303	21,428
Total population	18	55	128	291	664	1,522	3,505	8,077	18,606	42,856

Note: Yearly growth rate (λ) = 2.30; generation time = 2.4 years.

Table 18.5. Stable-Age Distributions for 3 Demographic Models of the Captive Black-Footed Ferret Population

Age Class	Optimistic (%)	Realistic (%)	Pessimistic (%)
0	67.7	62.5	47.1
1	19.0	19.9	20.0
2	7.8	9.4	12.5
3	3.2	4.4	7.8
4	1.3	2.1	4.9
5	0.5	1.0	3.0
6	0.2	0.5	1.9
7	0.1	0.2	1.2
8	<0.1	0.1	0.8
9	<0.1	<0.1	0.5
10	<0.1	<0.1	0.2

These percentages can be used to establish ranges for effective population sizes of black-footed ferrets. If breeding sex ratios are equal and variation in family size is approximately Poisson-distributed, then the ratio of the effective size to the number of breeders is 1 : 1. Under these assumptions, therefore, the proportions given in Table 18.5 and those above establish a range of potential N_e/N ratios. These ratios ($N_e/N = 0.3$ to 0.6) are typical of many captive populations.

Soule et al. (1986) discuss the maintenance of heterozygosity in relation to growth rate, generation length, and captive carrying capacity of the population. They recommend as an initial goal that captive populations be managed to retain 90% of the original heterozygosity of the founding population over a 200-year period. If the effective founder size, growth rate, and generation length of a population are known, then the captive carrying capacity necessary to maintain 90% of the original heterozygosity can be calculated.

Before these calculations can be completed, however, the effective size of the founder population needs to be estimated. Although 18 ferrets have been captured, not all can be considered founders since some are known to be related to others. Figure 18.2 illustrates the tentative pedigree for the captured animals. Of the 29 animals shown, only 12 can be considered founders of the living animals. However, not all of them have full genetic representation in the living population. For example, only 50% of the alleles of founder 653–54 are represented. Table 18.6 shows the probability distribution of founder alleles in the living population. Additionally, if the probability of both alleles being present is less that 0.9, it was assumed that only 1 allele was present.

Table 18.6. Probability That Founder's Alleles for Any Given Locus Are Present in the Living Population of Black-Footed Ferrets

Founder[a]	No. of Alleles Present		
	Both	Only One	Neither
Molly (F)	1.0	0	0
653–54 (F)	0	1.0	0
640–41 (M)	0.5	0.5	0
Willa (F)	1.0	0	0
Emma (F)	1.0	0	0
Annie (F)	1.0	0	0
Jenny (F)	1.0	0	0
Dean (M)	1.0	0	0
Mom (F)	1.0	0	0
Scarface (M)	1.0	0	0

[a]F = female; M = male.

These results indicate that only 18 of the possible 20 founder alleles at each locus (86%) are present in the living population. Founders 653–54 and 640–41 are not fully represented in the living ferrets, and their contribution is considered only half complete. The effective number of founders is therefore less than the total number of founders. In addition, the sex ratio of the founders is skewed. Seven of the 10 founders are female. This translates into an effective size of 8 (Kimura and Crow 1963). The effective size of the founders was therefore considered to be 8, taking into consideration both the unequal sex ratio and the loss of founder alleles.

Table 18.7 lists captive carrying capacities required to retain various levels of the original heterozygosity remaining after 200 years with different population N_e/N ratios for the 3 black-footed ferret population models discussed earlier.

For all N_e/N ratios around 0.5, captive carrying capacities over 1,500 are required to maintain 90% of the original variation for 200 years. A more reasonable expectation would be to preserve 80% of the original diversity. Using the realistic model and N_e/N ratio of 0.5, a captive carrying capacity of about 500 animals would be required to maintain 75 to 80% of the original heterozygosity. An alternative approach would be to retain 90% of the original diversity for a shorter period (50 years) since the program includes rapid reintroduction of animals back into the wild. Maintaining 90% of the diversity for 50 years also requires a carrying capacity of approximately 500 animals. Therefore, a captive

Table 18.7. Captive Carrying Capacities Necessary to Retain Selected Percentages of Original Heterozygosity

Model	N_e/N	Heterozygosity Remaining After 200 Years				
		70%	75%	80%	85%	90%
	0.2	940	1,440	2,878	64,937	—
	0.3	535	762	1,236	3,145	—
	0.4	381	514	853	1,638	—
	0.5	290	391	584	1,104	7,565
Pessimistic	0.6	235	317	463	853	3,665
	0.7	199	265	385	676	2,450
	0.8	172	228	330	566	1,909
	0.9	152	200	288	486	1,452
	1.0	137	179	255	427	1,206
	0.2	717	968	1,436	2,619	11,027
	0.3	466	605	881	1,491	4,392
	0.4	339	449	628	1,043	2,824
	0.5	271	352	492	817	2,052
Realistic	0.6	224	292	417	665	1,607
	0.7	189	251	352	561	1,332
	0.8	166	216	303	483	1,125
	0.9	148	193	271	433	986
	1.0	132	172	242	387	865
	0.2	676	888	1,268	2,040	5,125
	0.3	445	584	817	1,322	3,037
	0.4	332	427	597	966	2,172
	0.5	264	339	474	767	1,690
Optimistic	0.6	218	287	393	622	1,407
	0.7	139	243	339	537	1,190
	0.8	121	214	293	464	1,028
	0.9	108	108	264	409	907
	1.0	97	97	233	369	800

Note: Results are shown for the 3 models examined: pessimistic (yearly growth rate [λ] = 138%); realistic (λ = 212%); and optimistic (λ = 231%).

carrying capacity of 500 animals (effective size of 250) is suggested for the captive black-footed ferret population. These 500 ferrets would be distributed over the stable age distribution shown in Table 18.3 and would include approximately 200 breeding animals (for the realistic model) during the birth season. Again, these estimates are open for revision as additional information is obtained from the captive population.

Once the captive population has reached the carrying capacity, only

40 breeders per year are required to maintain a stationary, stable captive population. Young produced by additional breeders will be used to supply the reintroduction program. Demographic and genetic management should be implemented to maximize the effective size of the population while supplying the needs of both the reintroduction program and the captive population.

References

Allendorf, F. W., and R. F. Leary. 1986. Heterozygosity and fitness in natural populations of animals. In *Conservation biology: The science of scarcity and diversity*, ed. M. E. Soule, 57–76. Sunderland, Mass.: Sinauer Assoc.

Ballou, J. D. 1984. Strategies for maintaining genetic diversity in captive populations through reproductive technology. *Zoo Biology* 3:311–23.

Dobson, A. P., and R. M. May 1986. Disease and conservation. In *Conservation biology*, ed. M. E. Soule, 345–66. Sunderland, Mass.: Sinauer Assoc.

Foose, T. J. 1983. The relevance of captive propagation to the conservation of biotic diversity. In *Genetics and Conservation*, ed. C. M. Schonewald-Cox, S. M. Chambers, B. MacBryde, and L. Thomas, 374–401. Menlo Park, Calif.: Benjamin/Cummings.

Forrest, S. C., D. E. Biggins, L. Richardson, T. W. Clark, T. M. Campbell III, K. A. Fagerstone, and E. T. Thorne. 1988. Population attributes for the black-footed ferret at Meeteetse, Wyoming, 1981–1985. *J. Mammalogy* 69:261–73.

Frankham, R., H. Hemmer, O. A. Ryder, E. G. Cothran, M. E. Soule, N. D. Murray, M. Snyder. 1986. Selection in captive populations. *Zoo Biology* 5:127–38.

Gilpin, M. E., and M. E. Soule. 1986. Minimum viable populations: Processes of species extinction. In *Conservation biology: The science of scarcity and diversity*, ed. M. E. Soule, 19–34. Sunderland, Mass.: Sinauer Assoc.

Goodman, D. 1980. Demographic intervention for closely managed populations. In *Conservation biology*, ed. M. E. Soule and B. A. Wilcox, 171–96. Sunderland, Mass.: Sinauer Assoc.

Hedrick, P. W., P. F. Brussard, F. W. Allendorf, J. A. Beardmore, and S. Orzack. 1986. Protein variation, fitness and captive propagation. *Zoo Biology* 5:91–100.

Johansson, I. 1961. Studies on the genetics of ranch bred mink. I. The results of an inbreeding experiment. *Z. Tierzuecht. Zuechtungs-Biol.* 72:293–97.

Kimura, M., and J. F. Crow. 1963. The measurement of effective population number. *Evolution* 17:279–88.

Lacy, R. C. 1987. Loss of genetic diversity from managed populations: Interaction effects of drift, mutation, immigration, selection and population subdivision. *Conservation Biology.* 1:143–58.

MacCluer, J. W., J. L. VandeBerg, B. Read, and O. A. Ryder. 1986. Pedigree analysis by computer simulation. *Zoo Biology* 5:147–60.

Soule, M., M. Gilpin, W. Conway, and T. Foose. 1986. The millennium ark: How long a voyage, how many staterooms, how many passengers? *Zoo Biology* 5:101–14.

Wayne, R. K., L. Forman, A. K. Newman, J. M. Simonson, and S. J. O'Brien. 1986. Genetic monitors of zoo populations: Morphological and electrophoretic assays. *Zoo Biology* 5:215–32.

19

LYNN A. MAGUIRE

Managing Black-Footed Ferret Populations under Uncertainty: Capture and Release Decisions

Uncertainty is a pervasive feature of black-footed ferret management decisions. Because of the small remaining population, stochastic fluctuations in mortality and reproduction heavily influence the dynamics of both wild and captive populations. Despite 4 years of intensive field study of the Meeteetse population (Clark et al. 1986), significant gaps remain in understanding black-footed ferret reproductive physiology, behavior, and ecology. Management of the remaining wild and captive ferrets is affected, therefore, by uncertainty because of stochastic events in small populations and ignorance about ferret biology. Although ongoing research programs will reduce the latter, decision making under uncertainty will dominate black-footed ferret management for the foreseeable future.

Methods developed for analyzing business decisions under uncertainty have recently been applied to problems of endangered species (Maguire 1986, 1987; Maguire et al. 1987), including black-footed ferret management where prairie dog populations are being disrupted by sylvatic plague (Maguire and Clark 1985) and black-footed ferret management in Montana (Maguire et al. 1988). The purpose of this chapter is twofold: (1) to illustrate the application of formal methods for analyzing decisions under uncertainty to black-footed ferret management, using an analysis of capture and release decisions as an example; and (2) to draw conclusions from the analysis about research and management strategies concerning capture and release.

Previous analyses of black-footed ferret population dynamics (Wyoming Game and Fish 1986) and census results from the summer of 1986 (Oakleaf, telephone conversation, June 1986) led to recommendations to capture up to 14 additional ferrets from the wild population at Meeteetse for captive breeding. Despite widespread recognition that the remaining wild population at Meeteetse was too sparse to persist for long, the decision to capture essentially all the remaining wild ferrets was made with understandable reluctance. Not only would this action extinguish the wild population prematurely, it would also preclude further studies of wild ferrets that could facilitate both captive management and subsequent release of captive-bred ferrets. Some biologists (Wyoming Game and Fish 1986) have suggested that the presence of even a few wild ferrets might, through unspecified mechanisms, enhance the success of release efforts. All these concerns raise the question: should a remnant wild population be left behind, either by not capturing all of the ferrets in the wild or by releasing, more or less immediately, some previously captured ferrets; or should as many ferrets as possible be added to the captive-breeding program?

Related questions arise about the strategy for releasing captive-bred ferrets to reestablish a viable wild population at Meeteetse or elsewhere. At what level of productivity in the captive population should releases begin? How many animals should be released? Is it worthwhile to accelerate the release program, releasing animals at lower levels of captive productivity, in order to make releases while some wild ferrets remain? The release strategy should maximize the chances of successfully reestablishing a wild population, but minimize adverse impacts on productivity of the captive population, since repeated releases will be required and also because the threat of disease or other catastrophe, such as decimated the Meeteetse population recently, cannot be eliminated in the future.

The questions addressed by this decision analysis are then: Should any black-footed ferrets be left in the wild now? At what level of captive productivity from the captive population should releases of captive-bred ferrets begin? Should releases begin at lower levels of captive productivity in order to take advantage of any interaction between the wild remnant and released ferrets that may enhance release success? These questions will be examined using formal methods for analyzing decisions under uncertainty, as described by Raiffa (1968) and Behn and Vaupel (1982).

Formulating the Decision Analysis

In analyzing the questions posed in the introduction, I will assume that the overall objective of black-footed ferret management is to maximize the probability of species survival over the next 5 to 10 years, with both captive and wild populations contributing to this goal. The capture and release decisions outlined above, as well as the sources of uncertainty affecting the outcome of these decisions, can be organized in a decision tree (Fig. 19.1), in which the squares indicate decision points, the circles indicate chance events, and the chronological sequence is from left to right.

The most immediate decision is how many of the remaining wild ferrets to bring into captivity; the alternatives considered here are "capture all" (recognizing that it may be impossible to capture every last one) and "leave some" (meaning leave the 10 or so wild ferrets remaining after 1 or 2 adult males are captured and added to the previously captured 6). The outcome of each action in terms of both wild- and captive-population status depends on 2 uncertain events: (1) whether or not the captive population achieves some specified level of productivity, X (perhaps defined as the number of surplus juveniles produced per year), at a specific time, t (within the next 5 to 10 years); and (2) whether or not a reproductively active remnant population survives in the wild at time t. I assume that a decision to capture all precludes survival of a remnant to time t. The position of the chance nodes for these random events to the right of the capture-decision node indicates that the capture decision must be made in the face of uncertainty about future captive productivity and remnant survival. Although uncertainty about

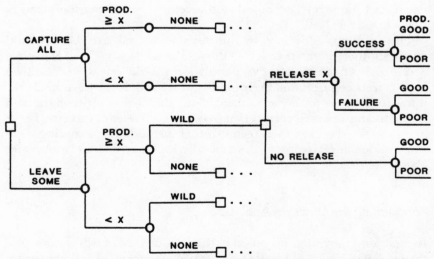

Figure 19.1. Decision tree for black-footed ferret capture and release decisions. Squares are decision nodes; circles are chance nodes. Release-decision node shown only once in detail; same subtree follows each ellipsis. X = target level of captive productivity (prod.).

these events could perhaps be reduced through modeling and experimental research on ferret population dynamics, some uncertainty will remain due to the stochastic behavior of small populations.

The subsequent decision represented in the tree (Fig. 19.1) is whether or not to release X captive-bred ferrets into the wild at time t. The location of the release-decision node to the right of the "captive productivity" and "wild remnant" nodes shows that the release decision will be made with knowledge of the status of the captive and wild populations at time t. Thus, it will be possible to tailor the release decision to each of the 4 possible population situations at time t: (1) captive productivity exceeds X, no wild remnant; (2) captive productivity less that X, no wild remnant; (3) captive productivity exceeds X, wild remnant survives; and (4) captive productivity less than X, wild remnant survives. The outcome of the decision to release or not in terms of wild and captive population status depends on: (1) the success or failure of the release to reestablish a reproducing population of black-footed ferrets in the wild (recognizing that assuring success will require repeated releases in subsequent years); and (2) captive productivity subsequent to the release decision, where "good" means productivity as high as before the release decision and "poor" means lower productivity. Captive productivity may be impaired by the harvest of animals for release to the wild; the level of impairment may depend on the level of captive productivity when the release decision was made, particularly when releasing X animals requires dipping into the reproductive capital of the captive population. (In approximating the continuous variables of captive productivity before and after the release decision with dichotomous choices [greater than or less than X, and good or poor], I am ignoring both age structure of the captive and released ferrets and the potential absurdity of releasing X ferrets when fewer than X captives are available.)

The capture decision must be made now, but its success may depend on later release decisions, which will be made under conditions that are presently unknown. Decision analysis provides a framework for this type of sequential decision making under uncertainty. Using this analysis, the capture alternative with the best chance of success, given current guesses about future conditions in the wild and captive populations, can be selected.

Estimating Probabilities of Chance Events and Values of Outcomes

Making a quantitative analysis of the best capture and release decisions requires estimates of the probabilities of the chance events that affect outcomes and of the values of the outcomes of each combination of management action and chance event. These probabilities and values may be estimated by stochastic simulation of both captive and wild population dynamics (models developed by Wyoming Game and Fish 1986; Harris et al., chap. 6, this volume; Groves and Clark 1986); by structured questioning to elicit expert opinion from ferret biologists and managers (Raiffa 1968; Behn and Vaupel 1982); and by a combination of these methods. For current purposes, I have assigned plausible estimates from my own experience and then analyzed the sensitivity of the results to deviations from these estimates.

On average, the productivity of the captive population will increase through time (Fig. 19.2a). The probability that captive productivity exceeds a specified level, X, also increases through time (Fig. 19.2b), approaching an asymptotic value less than 1 (since some realizations of the captive dynamics will lead to extinction). The probability that captive productivity exceeds X at time t depends on X, t, and the initial size of the captive population. For the black-footed ferret, a decision to capture all will result in an initial captive population of about 17 animals, versus about 7 if no additional captures are made, other than the adult males mentioned above (Fig. 19.2b). Captive productivity will increase faster

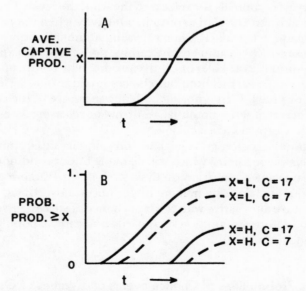

Figure 19.2. (A) Average (ave.) captive productivity (prod.) versus time. X = target level of productivity. (B) Probability (prob.) that captive productivity exceeds the target level X versus time for 2 initial captive population sizes (C) and for 2 target productivity levels (X), low (L) and high (H).

in the larger initial population. I assigned probabilities for 2 levels of captive productivity, low (L) and high (H), as in Table 19.1. High productivity may be thought of as about 40 animals, within the range of 30 plus per year that previous experience with mustelid releases suggests will be necessary for successful reestablishment (Richardson et al. 1986). Low productivity may be thought of as about 10 animals per year, a number that might be considered for release if it was urgent to begin releases early while some wild ferrets remain. For the probabilities in Table 19.1, the time t, at which the release decisions will be evaluated, may be thought of as about 4 years after the capture decision. Depending on captive breeding success in the next few years, however, the probabilities given in Table 19.1 might be appropriate sooner or later than that time.

If a decision is made to leave some ferrets in the wild, their numbers almost surely will decline with time, the rate of loss depending partly on the initial size of the remnant population (Fig. 19.3a). The probability that a reproductively active remnant survives at time t will also decline with time, again with the rate of decline depending on the initial remnant population size (Fig. 19.3b). If the initial remnant size is about 10, I estimated the probability of reproductively active ferrets remaining in 4 years (W) to be about 0.1.

The probability that release of X captive-bred ferrets will be successful in reestablishing the wild population depends partly on the number to be released, either L or H, with the larger number offering a higher probability of success. Success may also depend partly on the presence of a reproductively active remnant population at the release site. The mechanism by which the presence of a remnant might facilitate release success has not been specified. The effect could be direct, through cul-

Table 19.1. Probability That Captive Productivity at Time t Exceeds a Target Level X

	Captive Productivity X	
Captive Population	L	H
7	XLS = 0.1	XLS = 0.05
18	XAL = 0.5	XAL = 0.2

Note: Probability for 2 levels of productivity, L (low) and H (high); and for 2 initial captive population sizes: 7, if some ferrets are left in the wild; and 18, if all wild ferrets are captured. Parameters are defined in Appendix.

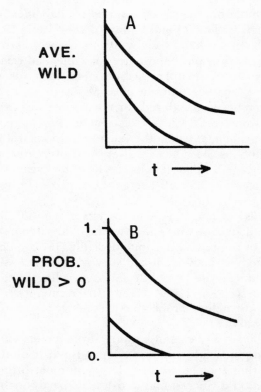

Figure 19.3. (A) Average (ave.) number of wild ferrets in remnant population versus time. (B) Probability (prob.) that a reproducing wild remnant population survives versus time.

tural learning (which does not seem very likely with animals as short-lived and widely dispersed as ferrets) or through breeding of released with wild ferrets; or the effect could be indirect, if additional study of wild ferrets improves techniques for release, or if the presence of wild ferrets enhances habitat protection at the release site. Any effect of a wild remnant on release success is likely to be reduced as the number of ferrets released becomes very large. Table 19.2 contains estimates of probability of release success if there is an impact of the wild remnant on success; if there is no such impact, the probability of success will be that estimated in the absence of a wild remnant (0.05 for *L* and 0.2 for *H*), whether a wild remnant survives or not.

The probability that captive productivity after the release decision will be unimpaired (that is, as high as before) depends on the number of captive-bred ferrets released (0, *L*, or *H*) and on the level of captive

Table 19.2.   Probability That Release of X
Captive-Bred Ferrets at Time t Will Be Successful
in Reestablishing the Wild Population

Wild Remnant	No. Released X	
	L	H
Present	$SW = 0.1$	$SW = 0.3$
Absent	$SNW = 0.05$	$SNW = 0.2$

Note: Probability for 2 release group sizes, L (low) and H
(high); and for remnant wild ferret population present or
absent. Parameters are defined in Appendix.

productivity before the release decision (greater than or less than X). Of
course, it is possible for captive productivity to be so high that releasing X
ferrets has no impact on future productivity; and it is possible that there
may be no correlation between captive productivity before and after the
release decision. However, since the release decision will ordinarily arise
where captive productivity is low enough to produce uncertainty about
the impact of releases on future productivity, and where higher productivity before the release decision is likely to be correlated with higher
productivity after the decision, the probabilities will generally be or-
dered as shown in Table 19.3, where $XXN > XXR$; $XXN > XN$; $XN > XR$;
and $XXR > XR$. (See Appendix for parameter definitions.)

In order to decide among the alternative management strategies, each
possible outcome of the actions and chance events in the decision tree
must be evaluated according to criteria that reflect the objective of ferret

Table 19.3.   Probability That Captive Productivity Following the Release
Decision Is at Least as High as It Was at Time t ("Good")

Captive Productivity	No. Released X			
	L		H	
	L	0	H	0
>X	$XXR = 0.6$	$XXN = 0.9$	$XXR = 0.7$	$XXN = 0.9$
<X	$XR = 0.2$	$XN = 0.4$	$XR = 0.3$	$XN = 0.5$

Note: Probability for release group sizes of L (low), H (high), and 0; and for a level of
captive productivity at time t either greater than or less than X (where X = L or X = H).
Parameters are defined in Appendix.

management: to maximize species survival using both captive and wild populations. In qualitative terms, the possible outcomes for the wild population, after the release decision has been made, are: (1) a wild population (which may or may not include a wild remnant) has been reestablished by successful release of captive-bred ferrets; (2) a reproductively active wild remnant survives; or (3) no wild ferrets remain. The possible outcomes for the captive population are: (1) captive productivity is good; or (2) captive productivity is poor. All 6 combinations of wild and captive outcomes are possible, as shown in Table 19.4. The values assigned to each of the 6 outcomes should reflect the relative preferences of ferret managers, incorporating appropriate weights for well-being of the captive population compared to that of the wild. These relative preferences, or utilities, can be elicited using a structured series of questions, as described by Raiffa (1968) or Behn and Vaupel (1982).

For this analysis, I have assigned the values shown in Table 19.4. The best outcome, reestablished wild population and good captive productivity, receives a preference rating of 1; the worst outcome, no wild ferrets and poor captive productivity, receives a rating of 0. The assignment of values to the remaining 4 outcomes assumes that the presence of a wild remnant is not much better than no wild ferrets (value about 0.1); that the relative value of a reestablished wild population is somewhat higher (adds 0.5 to 0.7 to overall value) than that of good versus poor captive productivity (adds 0.3 to 0.5 to overall value); that the relative value of the wild population is higher when captive productivity is poor (0.7) than when captive productivity is good (0.5); and that the relative value of good captive productivity is higher when the wild population exists as a remnant or not at all (0.5) than when the wild population has been successfully reestablished (0.3). These relationships reflect a bal-

Table 19.4.   Values Assigned to Possible Outcomes, Expressing Relative Preference for Healthy Wild versus Healthy Captive Population

Wild Status	Captive Productivity	Value
Reestablished	Good	1.0
Reestablished	Poor	0.7
Remnant	Good	0.6
Remnant	Poor	0.1
None	Good	0.5
None	Poor	0.0

Note: Wild status and captive productivity assessed after the release decision.

ance between wild and captive populations in which survival in the wild is the ultimate concern, but maintaining a healthy captive population is necessary to assure wild survival.

Stochastic Simulation and Goals for This Decision Analysis

Stochastic simulation of captive and wild population dynamics offers one means of estimating some of the probabilities used in the decision analysis. In theory, an analysis of potential capture and release strategies using a combination of stochastic simulation, dynamic programming, and utility analysis could provide completely general answers to questions concerning how many ferrets should be left in the wild, at what level of captive productivity should release of captive-bred ferrets begin, and what numbers of ferrets should be released. In practice, such an analysis would be so unwieldy and time-consuming as to be virtually useless for management purposes. Instead, the analytic framework outlined above can be used to direct modeling and experimental efforts so that a more limited set of simulation studies, in conjunction with decision analysis, can provide reliable guidance for management decisions.

The goals for this decision analysis are then: (1) to identify the best capture and release decision strategy for particular parameter estimates (L and H); (2) to understand how the components of the decision problem influence capture and release decisions; (3) to direct future simulation and experimental investigations to the most critical components for determining the best decision strategy; and (4) where possible, to draw general conclusions from the decision analysis alone about the best capture and release decisions.

Note that the *capture all* decision can be made only once, whereas the decision to *leave some* could be reevaluated at successive times t. Similarly, the release decision can be reevaluated at successive times t, up to the point at which successful releases have been made. This chapter will focus on the first consideration of the capture and release decisions, with a few comments on how these may influence subsequent reconsideration; however, the same analysis techniques can be used for subsequent decisions as well.

Optimal Decision Strategies

According to the principles of decision theory, the best decision under uncertainty is the one that maximizes expected utility (or relative preference) (Raiffa 1968; Behn and Vaupel 1982). To determine the best strategy from a decision tree, such as the one in Figure 19.1, the possible outcomes of each action or event branch in the tree (listed at the ends of the branches of the more detailed trees in Figs. 19.4 and 19.5) are weighted by their probabilities of occurrence (listed on the random events branches in Figs. 19.4 and 19.5) to form an expected value at each of the chance nodes on the tree. At each decision node, the branch with the higher expected value is selected. Working from right to left through the tree gives the best sequence of decisions for a particular set of estimated probabilities and values.

For parameter set L in this analysis, the optimal sequence of decisions is to capture all wild ferrets initially and not to release L ferrets at time t under any of the 4 conditions of wild status and captive productivity (expected value = 0.325). For parameter set H, the best sequence of decisions is to capture all wild ferrets initially and to release H ferrets under all 4 conditions subsequently (expected value = 0.316). Although these results are of some interest, even more important is an understanding of how these decision strategies might change if the parameters were somewhat different from those estimated here. For example, the decisions to capture all and to make no release are quite clear-cut for the L parameter set. But for the H parameter set, the expected values for

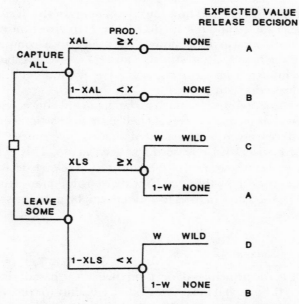

Figure 19.4. Detailed tree for the capture decision. Parameters defined in Appendix.

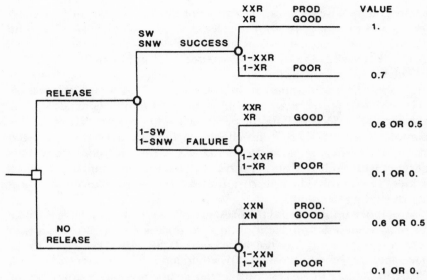

Figure 19.5. Detailed tree for the release decision. Parameters defined in Appendix. Higher values associated with outcomes are when wild remnant survives (0.6 and 0.1); lower values are when no remnant survives (0.5 and 0). (Prod. = productivity)

release and *no release* when no wild remnant survives, and for *capture all* versus *leave some*, differ by only about 0.1, suggesting that modest changes in the parameter estimates may change these decisions. Sensitivity analysis of both the capture and release decisions helps to develop an understanding of the components that influence decision strategy and guides further research to develop confidence in the recommended strategies.

Sensitivity Analysis of the Release Decision

The sensitivity analysis will be divided into 2 stages: (1) the release decision and (2) the capture decision. Sensitivity analysis can be used both to ask how a decision might change in response to specific changes in one or more parameters and to elaborate the optimal decision strategies for a range of possible parameter values. I will emphasize the latter approach here as a means of directing further research.

The parameters influencing the release decision are: (1) the probability of successful release, with and without a wild remnant (*SW* and *SNW*); (2) the probability that captive productivity is good or poor following the release decision, as a function of the number of ferrets released and productivity prior to release (*XXR, XXN, XR,* and *XN*); and (3) the values assigned to outcomes, particularly the relative weights assigned to well-being of the captive and wild populations (Fig. 19.5).

The first question addressed by the sensitivity analysis was, How does the impact of a wild remnant on release success influence the release decision? If the presence of a wild remnant does not enhance release success ($SW = SNW = 0.05$ for parameter set *L; SW = SNW = 0.2* for parameter set *H*), there is no change in the optimal strategies listed above. However, for parameter set *H* the release decision is now borderline, with "release" and "no release" having very similar expected values. There would be little cost, under these circumstances, to choosing the wrong decision.

How much impact would the presence of a wild remnant have to make on release success to make it optimal to release *L* ferrets when a wild remnant is present, but not otherwise? If all other values in the *L* parameter set remain constant, *SW* would have to be greater than about 0.31 (probability of success with a wild remnant about 6 times higher than without). Since it seems unlikely that the presence of a few remaining wild ferrets could enhance release success to such a degree, this result suggests that releasing *L* ferrets is not likely to be a good decision. This illustrates one of the benefits of analyzing sensitivity in a decision problem: identifying those decision points where a parameter value can change substantially without affecting the optimal decision. It is difficult to anticipate where these points occur from intuitive examination of the problem.

The response of the release decision strategy for parameter set *L* to changes in *SW* and *SNW* is summarized in Figure 19.6. Most of the behavior illustrated in this figure conforms to intuition: for example, releases should be made when probabilities of success are high, but not when low. Other responses are counterintuitive: for some *SW* and *SNW* combinations, the optimal strategy is to release *L* ferrets when productivity is low, but not when productivity is high. Upon reflection, these recommendations make sense in light of the relative values assigned to captive and wild population status; but, again, it would be difficult to foresee such conclusions from an intuitive analysis of the problem.

In structuring the decision problem, I assumed that if productivity after the release decision was positively correlated with productivity before, and if harvesting captive-bred animals for release adversely affected future captive productivity, there would be some parameter combinations where the optimal strategy is to release *X* ferrets when captive productivity exceeds *X*, but not when it falls short of the target level.

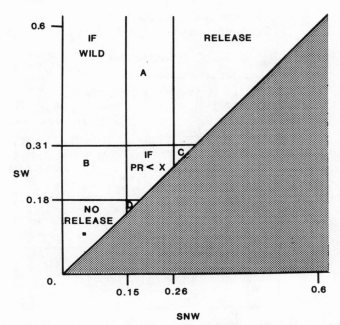

Figure 19.6. Sensitivity of release decision to changes in probability of successful release, with (*SW*) and without (*SNW*) wild remnant, for parameter set *L; PR* = captive productivity at time *t. A* = release if remnant present, or if *PR* < *X* and remnant absent; *B* = release if *PR* < *X* and remnant present; *C* = release if *PR* < *X*, or *PR* > *X* and remnant absent; *D* = release if *PR* < *X* and wild absent. Dot marks nominal values for *SW* and *SNW*. Shaded area indicates infeasible parameter combinations.

Analysis of the sensitivity of the release decision to changes in *XXR* and *XR* for parameter set *H* (Fig. 19.7) shows that this is the case when *XXR* is greater than about 0.67 and when *XR* is less than about 0.16. Figure 19.7 also shows that the nominal values of *XXR* and *XR* for parameter set *H* are close to the boundaries where the optimal decision is to release *H* ferrets under all conditions. Relatively small decreases in *XXR* or *XR* would change the optimal decision strategy. Most of the strategies illustrated in Figure 19.7 are intuitively understandable, that is, release under all conditions when neither release of *H* ferrets nor low productivity prior to the release decision adversely affects future captive productivity (*XXR* and *XR* both high). But, again, some decision strategies at first seem counterintuitive, that is, for some combinations of *XXR* and *XR*, it is best to release *H* ferrets only when captive productivity is less than *X*.

The response of the release decision to *XXR* or *XR, XXN* or *XN*, and *SW* or *SNW* is summarized in Figure 19.8. For parameter combinations to the upper left of the *SW* or *SNW* line, the optimal decision is to release

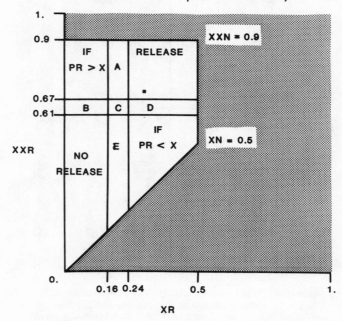

Figure 19.7. Sensitivity of release decision to changes in probability that future captive productivity is "good," for current captive productivity $(PR) > X (XXR)$ and $< X (XR)$, for parameter set H. A = release when $PR > X$, or $PR < X$ and remnant present; B = release when $PR > X$ and remnant present; C = release when remnant present; D = release when $PR < X$, or $PR > X$ and remnant absent; E = release when $PR < X$ and remnant present. Dot marks nominal values for XXR, XR. Shaded area indicates infeasible parameter combinations.

X ferrets; to the lower right of the line, the optimal decision is no release. The higher the probability of successful release (SW or SNW), the more combinations of XXR, XR, XXN, XN there are for which the optimal decision is to release X ferrets, as expected. When SW is greater than about 0.83, or when SNW is greater than about 0.7, release is the optimal decision regardless of the values of XXR, XR, XXN, and XN. The break-

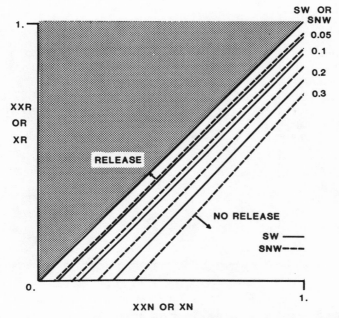

Figure 19.8. Summary of sensitivity of release decision to parameters describing captive productivity after release decision (*XXR, XXN, XR, XN*) and probability of successful release (*SW, SNW*). Release is optimal for parameter combinations to upper left of the appropriate break-even line; no release is optimal for parameter combinations to the lower right. Shaded area indicates infeasible parameter combinations. In addition, *SW* > or = *SNW*.

even lines for these *SW* and *SNW* values pass through the point (1,0). (Note that *SW* will always be greater than or equal to *SNW*, unless the presence of a wild remnant decreases, rather than increases, the probability of release success.)

The release decision is also affected by the relative importance assigned to a healthy captive population versus a healthy wild population. The parameter space for relative weights of captive and wild status falls on the line $x + y = 1$, because the outcome with reestablished wild population and good captive productivity has been assigned a preference value of 1 (Table 19.4). When the relative value of the captive population is high, no releases will be made. For the *L* parameter set, releases of *L* ferrets are possible only when the relative value of a reestablished population is at least 150% that of good captive productivity. For the *H* parameter set, releases can be made when the value of the wild population is at least that of the captive. Figure 19.9 summarizes response of the release decision to changes in relative values of wild and

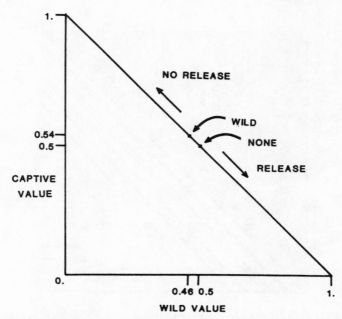

Figure 19.9. Sensitivity of release decision to relative values assigned to health of captive and wild populations, for parameter set *H;* relative value of wild remnant = 0.1. Optimal decision is to release for points on line above *wild,* not to release for points below *none,* and to release only when wild remnant present for points between *wild* and *none.*

captive status for parameter set *H,* where the remnant wild population has a value (*R*) of 0.1. When the relative values of wild and captive status are to the upper left of the point marked *wild,* no releases of *H* ferrets are made under any conditions; between the points marked *wild* and *none,* releases are made when a wild remnant is present, but not otherwise; to the lower right of the point *none, H* ferrets would be released under any conditions. For higher relative values assigned to the remnant wild population, the break-even points for releases generally shift to the lower right on the line, making releases less likely. Most of these results conform to intuition, since when captive productivity has high value relative to the wild population, it makes sense to keep all ferrets in captivity, rather than jeopardize captive production by harvesting ferrets for release. Conversely, when reestablishing the wild population has higher relative value, risking the productivity of the captive population may be justifiable to achieve improved status in the wild.

The sensitivity analysis of the release decision has shown that, in general, the optimal decision strategies conform to intuition, that is, releases should be made when captive productivity is high, when probability of success is high, and when the relative value of the wild population is high. There are many plausible parameter combinations, however, for which a counterintuitive decision is optimal, such as releasing when wild ferrets are absent, but not when they are present. These counterintuitive results suggest using simulation studies, or other methods, to establish which region of the parameter space (see Figs. 19.6–19.9) applies before making the release decision. The sensitivity analyses show which decisions are unlikely to be affected by modest changes from the nominal parameter values, for example, the impact of a wild remnant on success in releasing L ferrets is not likely to be large enough to justify release. They also show which decisions are more likely to change, that is, release of H ferrets would no longer be optimal for relatively small decreases in several parameters.

Sensitivity Analysis of the Capture Decision

The analysis of the optimal decision strategy for both the L and the H parameter sets showed that the best initial decision is to capture all the remaining wild ferrets. For the L parameter set, this initial decision is followed by a decision not to release L ferrets at time t; whereas, for the H parameter set, the subsequent decision is to release H ferrets at time t. Under what conditions would the optimal capture decision include leaving some ferrets in the wild; in particular, will it ever be optimal to leave some ferrets when the subsequent decision is to make no releases? Intuitively, I expected the balance between the benefit of additional captures in enhancing captive productivity (XAL versus XLS) and the benefit of a wild remnant in enhancing release success (SW versus SNW) to determine the optimal capture strategy.

The capture decision may depend on the expected values of the outcomes of the release decision analyzed above. These expected values are represented as A, B, C, and D on the simplified tree for the capture decision in Figure 19.4. Using the probabilities to obtain the expected values at each chance node produces the following expression for the expected value of "capture all" versus "leave some":

$$XAL <?> XLS \left[1 + W (C - A + B - D)/(A - B)\right] \\ + W (D - B)/(A - B) \qquad (19.1)$$

The differences $(A - B)$ and $(C - D)$ can be attributed to correlation between captive productivity before and after the release decision. If there is no correlation ($XXR = XR$ and $XXN = XN$), then $A = B$ and $C = D$; and Equation 19.1 becomes:

$$XAL < ? > XLS + \infty \qquad (19.2)$$

provided that the probability of a wild remnant surviving (W) is greater than 0, and provided that there is some value to having a wild remnant survive ($R > 0$) or some beneficial impact of the wild remnant on release success ($SW > SNW$), which ensures that $D > B$. In this case, it is always optimal to leave some ferrets in the wild, regardless of the values of XAL, XLS, and W. This makes good intuitive sense, because if current and future productivity are not positively correlated, there is no need to make additional captures in order to improve the likelihood that captive productivity at time t will exceed X.

For parameter combinations where no releases are made under any conditions, $(C - A) = R$ and $(D - B) = R$, and Equation 19.1 becomes:

$$XAL < ? > XLS + W\,[R/(A - B)] \qquad (19.3)$$

A line with slope 1 and intercept $W\,[R/(A - B)]$ describes the break-even conditions to the upper left of which the optimal decision is to capture all, and to the lower right of which the optimal decision is to leave some ferrets in the wild, as shown in Figure 19.10 for parameter set L and various levels of W. For higher probability of remnant survival (W), higher value of the remnant population (R), or lower correlation of current and future captive productivity ($A - B$), the intercept of the break-even line is larger, and it is less likely that XAL will exceed XLS by enough to justify capturing all. However, for the nominal values of parameter set L ($XAL = 0.5$, $XLS = 0.1$, $W = 0.1$), the optimal decision is clearly to capture all, and this conclusion is not likely to be affected by even fairly large changes in these parameters.

For parameter combinations where releases will be made under all conditions, and where there is no beneficial impact of a wild remnant on release success ($SW = SNW$), the break-even line between the 2 capture alternatives is again given by Equation 19.2. In this case, presence or absence of a wild remnant is irrelevant to the expected value of the outcome except for the value assigned to remnant survival (R).

For any parameter combinations, the break-even line between capture alternatives may be represented by Equation 19.1. Figure 19.11 shows optimal capture strategies for parameter set H, where H ferrets are to be released under all conditions, for various levels of W, XAL, and XLS. The differences ($A - B$) and ($C - D$) may be attributed to correlation between current and future captive productivity. The differences ($C - A$) and ($D - B$) may be attributed partly to the value assigned to remnant survival (R) and partly to any beneficial impact of remnant survival on release success (SW versus SNW). The differences ($A - B$)

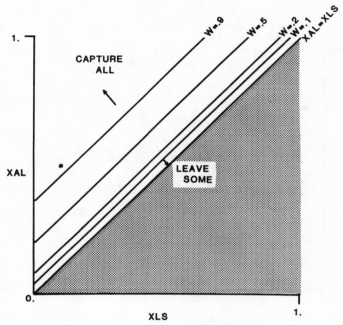

Figure 19.10. Sensitivity of capture decision to probability that captive productivity exceeds X at time t if all wild ferrets captured (XAL) or if some left in wild (XLS), for parameter set L (where release decision is no releases) and various probabilities of wild remnant survival (W). Optimal decision is to capture all for parameter combinations to the upper left of the appropriate break-even line and to leave some wild ferrets for combinations to the lower right. Dot marks nominal values. Shaded area indicates infeasible parameter combinations.

and $(C - D)$ are affected similarly by changes in parameters, and the differences $(C - A)$ and $(D - B)$ are affected similarly by changes in parameters. Therefore, the numerator of the bracketed term in Equation 19.1 is likely to stay close to 0; it is then divided by a probably larger quantity $(A - B)$ and multiplied by a number less than 1 (W). For these reasons, the slope of the break-even line is unlikely to be far from 1. Therefore, it is the difference between XAL and XLS, rather than a ratio, that determines the optimal capture strategy. The probability of a wild remnant surviving (W) is important in determining the intercept of the break-even line, with a higher W making it more likely that leaving some wild ferrets is the optimal strategy, which makes good sense.

Discovering that the slope of the break-even line between capture alternatives is likely to be close to 1 helps direct efforts to estimate XAL and XLS. Simulation or other studies should concentrate on determin-

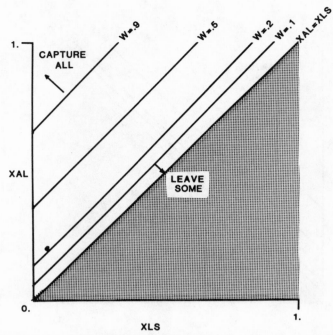

Figure 19.11. Sensitivity of capture decision to probability that captive productivity exceeds X at time t if all wild ferrets captured (XAL) or if some left in wild (XLS), for parameter set H (where release decision is to release H ferrets) and various probabilities of wild remnant survival (W). Optimal decision is to capture all for parameter combinations to the upper left of the appropriate break-even line and to leave some wild ferrets for combinations to the lower right. Dot marks nominal values. Shaded area indicates infeasible parameter combinations.

ing the difference between XAL and XLS, because it is the difference between them, rather than their absolute values, that determines the optimal capture strategy. This conclusion would be difficult to obtain from an intuitive examination of the problem.

It is also important to use simulation or other means to estimate W, which critically determines the intercept of the break-even lines. If W is low, as seems likely, the optimal decision is almost certainly to capture all. Considering the current sizes of wild and captive populations, it seems very likely that $XAL - XLS$ is large enough, and W small enough, to put the capture decision well within the *capture all* region of the parameter space.

As time goes on, both XAL and XLS will increase (Fig. 19.2b) and W will decrease (Fig. 19.3b). Increase in both XAL and XLS will move toward the upper right in the parameter space of graphs such as Figures

19.10 and 19.11. Since W is decreasing simultaneously, it is very unlikely to move with increasing time from the *capture all* to the *leave some* portion of the parameter space, although it is possible that the reverse could occur. If an analysis based on current assessment of captive and wild population dynamics suggests that capturing all is the best decision, the decision will probably not be regretted later. This is a reassuring result, because it is the *capture all* decision that cannot be undone later.

Conclusions from this Analysis

This decision analysis of capture and release decisions for black-footed ferret management has yielded several types of results. First, conclusions were drawn about the optimal decision sequence for particular parameter sets, L and H. Second, the analysis showed which decisions are unlikely to change without large changes in parameter values, for example, the decision not to release L ferrets. In addition, the analysis helped to (1) outline the relationships among parameters and decision strategies, pointing out both intuitive and counterintuitive relationships; (2) identify the critical values of parameters where the optimal decision changes from one alternative to the other; and (3) characterize which combinations of parameters are critical to decision making, such as the difference between XAL and XLS. Decision analysis can sometimes produce robust conclusions about decision strategies, even when information is very limited. For example, a decision to capture all remaining wild ferrets is probably optimal for a wide range of likely parameter values, offering some confidence to proceed with that strategy despite its irreversibility.

Decision analysis also helps guide simulation and experimental studies, directing efforts to understanding critical parameter relationships, such as the difference between XAL and XLS, and the level of W. In this case, limited simulation runs to establish the slope and intercept for the break-even line for the capture decision, followed by simulation studies aimed at determining $XAL - XLS$ and W, should be a fruitful approach. This is a more focused and manageable use of simulation than the complete investigation of capture and release strategies described earlier in this chapter. It takes advantage of decision analysis to identify regions of the parameter space over which a particular decision is optimal. In most cases, to make decisions with confidence, simulation need point only to the region that applies, not to a particular spot in that region.

Some of the benefits that can be expected from decision analysis are:

(1) specification of optimal strategies for particular parameter sets; (2) an understanding of the direction and magnitude of parameter changes that will change the optimal decision strategy: (3) some general conclusions about optimal strategy even where there is considerable uncertainty about the parameter estimates; and (4) guidance of further research.

Other Black-Footed Ferret Management Decisions under Uncertainty

Many other aspects of black-footed ferret management involve decision making under uncertainty. If additional populations of ferrets are found in the wild, decisions must be made about capture, translocation to other wild sites, addition to the existing captive population or initiation of a new captive-breeding effort, and intensity of management for populations left in the wild. These decisions will be made under uncertainty about wild and captive population dynamics, including demographic stochasticity, uncertain genetic relationships, catastrophic events in the wild and captive environment, and so on. Some of these questions have received preliminary analysis for the ferret program in the state of Montana (Maguire et al. 1988).

The urgent need for successful breeding of captive ferrets raises concerns about surgical and hormonal intervention in natural breeding as a means of facilitating reproduction. The decision to artificially manage captive breeding must be made with limited knowledge of ferret reproductive physiology and behavior, and with the risk that surgical or hormonal manipulation could injure or kill one of the few captive animals.

Management of the captive population will continue to involve decisions under uncertainty: at what point should the captive population be subdivided; what other facilities should attempt to rear black-footed ferrets; what patterns of exchange of animals or gametes should be maintained among subpopulations; should black-footed ferrets from other geographic areas be maintained separately; should Siberian ferrets be used in the black-footed ferret program?

Furthermore, decision analysis can be helpful in addressing sociopolitical, as well as scientific, uncertainties affecting black-footed ferret management. The decision to capture ferrets at Meeteetse has received intense public scrutiny, as has the preceding research program. In the above analysis of capture and release decisions, I have ignored the public relations aspects of capturing all the remaining black-footed ferrets. These are certainly important to the success of black-footed ferret management and can be considered in a decision-analysis framework by making public opinion one of the criteria used in evaluating possible outcomes.

In summary, the principles of decision making under uncertainty will continue to be useful for black-footed ferret management, whether applied informally or formally. Formal analysis offers many benefits by increasing confidence in recommended actions and by guiding the direction of future research.

Appendix: List of parameter definitions and nominal estimates:

A = expected value of outcome after release decision for captive productivity > X, no wild remnant branches.

B = expected value of outcome after release decision for captive productivity < X, no wild remnant branches.

C = expected value of outcome after release decision for captive productivity > X, wild remnant survives branch.

D = expected value of outcome after release decision for captive productivity < X, wild remnant survives branch.

PR = captive productivity at time t.

R = relative value assigned to survival of a wild remnant (0.1).

SNW = probability that release of X ferrets will be successful in reestablishing the wild population when no wild remnant survives (0.05 for $X = L$; 0.2 for $X = H$).

SW = probability that release of X ferrets will be successful in reestablishing the wild population when a wild remnant survives (0.1 for $X = L$; 0.3 for $X = H$).

W = probability that wild remnant with initial size about 10 survives at time t (0.1).

X = target captive-productivity level; and number of ferrets to be released, L (low) or H (high).

XAL = probability that captive productivity exceeds X at time t, given that all wild ferrets were captured, that is, an initial captive population size of about 17 (0.5 for $X = L$; 0.2 for $X = H$).

XLS = probability that captive productivity exceeds X at time t, given some ferrets left in wild, that is, an initial captive population size of about 7 (0.1 for $X = L$; 0.05 for $X = H$).

XN = probability that captive productivity following the release decision is at least as high as before, given that captive productivity at time $t < X$ and that no releases were made (0.4 for $X = L$; 0.5 for $X = H$).

XR = probability that captive productivity following the release decision is at least as high as before, given that captive productivity at time $t < X$ and that X ferrets were released (0.2 for $X = L$; 0.3 for $X = H$).

XXN = probability that captive productivity following the release decision is at least as high as before, given that captive productivity at time $t > X$ and that no releases were made (0.9 for both $X = L$ and $X = H$).

XXR = probability that captive productivity following the release decision is at least as high as before, given that captive productivity at time $t > X$ and that X ferrets were released (0.6 for $X = L$; 0.7 for $X = H$).

References

Behn, R. D., and J. W. Vaupel. 1982. *Quick analysis for busy decision makers*. New York: Basic Books.

Clark, T. W., et al. 1986. The black-footed ferret. *Great Basin Nat. Memoirs* 8:1–208.

Groves, C. R., and T. W. Clark. 1986. Determining minimum population size for recovery of the black-footed ferret. *Great Basin Nat. Memoirs* 8:150–59.

Maguire, L. A. 1986. Using decision analysis to manage endangered species populations. *J. Environ. Manag.* 22:345–60.

———. 1987. Decision analysis: A tool for tiger conservation and management. In *Tigers of the world*, ed. R. L. Tilson and U. S. Seal 475–86. Park Ridge, N.J.: Noyes.

Maguire, L. A., and T. W. Clark. 1985. A decision analysis of management options concerning plague in the Meeteetse black-footed ferret habitat. Report to the U.S. Fish and Wildlife Service, Denver Regional Office, July 15, 1985.

Maguire, L. A., T. W. Clark, R. Crete, J. Cada, C. Groves, M. L. Shaffer, and U. S. Seal. 1988. Black-footed ferret recovery in Montana: A decision analysis. *Wildl. Soc. Bull.* 16:111–20.

Maguire, L. A., U. S. Seal, and P. F. Brussard. 1987. Managing critically endangered species: The Sumatran rhino as a case study. In *Viable populations for conservation*, ed. M. E. Soule 141–58. Cambridge: Cambridge Univ. Press.

Raiffa, H. 1968. *Decision analysis*. Reading, Mass. Addison-Wesley.

Richardson, L., et al. 1986. Black-footed ferret recovery: Some reintroduction and captive breeding options. *Great Basin Nat. Memoirs* 8:169–84.

Wyoming Game and Fish Department. 1986. Recommendations for capture of black-footed ferrets to enhance the captive breeding program in 1986. Cheyenne, Wyo.: Wyoming Fish and Game Department.

Contributors

Rupert P. Amann
Animal Reproductive Laboratory
Colorado State University
Fort Collins, Colorado

Elaine Anderson
Denver Museum of Natural History
Denver, Colorado

Stanley Anderson
U.S. Fish and Wildlife Service
Cooperative Research Unit
University of Wyoming
Laramie, Wyoming

Robert W. Atherton
Department of Zoology
University of Wyoming
Laramie, Wyoming

Jonathan D. Ballou
Department of Zoological Research
National Zoological Park
Smithsonian Institution
Washington, D.C.

David W. Belitsky
Wyoming Game and Fish
  Department
Cody, Wyoming

Michael Bogan
U.S. Fish and Wildlife Service

National Ecology Research Center
Ft. Collins, Colorado

Richard A. Bowen
Animal Reproductive Laboratory
Colorado State University
Fort Collins, Colorado

Peter F. Brussard
Biology Department
Montana State University
Bozeman, Montana

Paul Budworth
Animal Reproductive Laboratory
Colorado State University
Fort Collins, Colorado

Warren Burgess
Department of Zoology
University of Wyoming
Laramie, Wyoming

Tim W. Clark
Northern Rockies Conservation
  Cooperative
Jackson, Wyoming; and
Chicago Zoological Society
Brookfield, Illinois

Brian P. Cole
U.S. Fish and Wildlife Service
Department of the Interior
Asheville, North Carolina

Patrick Curry
Department of Zoology
University of Wyoming
Laramie, Wyoming

Michael W. DonCarlos
Minnesota Zoological Garden
Apple Valley, Minnesota

Mary A. Eichelberger
Laboratory of Viral Carcinogenesis
National Cancer Institute
Frederick, Maryland

Nathan Flesness
International Species Information
  System
Minnesota Zoological Garden
Apple Valley, Minnesota

Thomas J. Foose
American Association of Zoological
  Parks and Aquariums
Minnesota Zoological Garden
Apple Valley, Minnesota

Michael Gilpin
Department of Biology
University of California
San Diego, California

Karen L. Goodrowe
Department of Animal Health
National Zoological Park
Smithsonian Institution
Washington, D.C.

Richard B. Harris
U.S. Fish and Wildlife Service
Fairbanks, Alaska

Joseph Herbert
Department of Anatomy

Cambridge University
Cambridge, England

Robert M. Kitchin
Department of Zoology
University of Wyoming
Laramie, Wyoming

Robert C. Lacy
Department of Conservation
  Biology
Brookfield Zoo
Brookfield, Illnois

Lynn A. Maguire
School of Forestry and
  Environmental Studies
Duke University
Durham, North Carolina

Janice S. Martenson
Laboratory of Viral Carcinogenesis
National Cancer Institute
Frederick, Maryland

Rodney A. Mead
Department of Biological Sciences
University of Idaho
Moscow, Idaho

Brian Miller
Department of Zoology
University of Wyoming
Laramie, Wyoming

Bruce D. Murphy
Department of Obstetrics and
  Gynecology
College of Medicine
University of Saskatchewan
Saskatoon, Saskatchewan, Canada

Robert Oakleaf
Wyoming Game and Fish

Department
Lander, Wyoming

Stephen J. O'Brien
Laboratory of Viral Carcinogenesis
National Cancer Institute
Frederick, Maryland

Ulysses S. Seal
Research Service
Veterans Administration Medical
   Center
Minneapolis, Minnesota

Mark L. Shaffer
U.S. Fish and Wildlife Service
Office of International Affairs
Washington, D.C.

Richard Slaughter
Department of Zoology
University of Wyoming
Laramie, Wyoming

Monte Straley
Department of Zoology
University of Wyoming
Laramie, Wyoming

E. Tom Thorne
Wyoming Game and Fish
   Department
Laramie, Wyoming

David E. Wildt
Department of Animal Health
National Zoological Park
Smithsonian Institution
Washington, D.C.

Frank Wright
c/o ISIS
Minnesota Zoo
Apple Valley, Minnesota

Index